SpringerBriefs in M

Extremophilic Bacteria

Series editors
Sonia M. Tiquia-Arashiro, Dearborn, USA
Melanie Mormile, Rolla, USA

More information about this series at http://www.springer.com/series/11917

Prasanti Babu · Anuj K. Chandel
Om V. Singh

Extremophiles and Their Applications in Medical Processes

Springer

Prasanti Babu
Om V. Singh
Division of Biological and Health Sciences
University of Pittsburgh
Bradford, PA
USA

Anuj K. Chandel
Department of Chemical Engineering
University of Arkansas
Fayetteville, AR
USA

ISSN 2191-5385
ISBN 978-3-319-12807-8
DOI 10.1007/978-3-319-12808-5

ISSN 2191-5393 (electronic)
ISBN 978-3-319-12808-5 (eBook)

Library of Congress Control Number: 2014953259

Springer Cham Heidelberg New York Dordrecht London

© The Author(s) 2015

This work is subject to copyright. All rights are reserved by the Publisher, whether the whole or part of the material is concerned, specifically the rights of translation, reprinting, reuse of illustrations, recitation, broadcasting, reproduction on microfilms or in any other physical way, and transmission or information storage and retrieval, electronic adaptation, computer software, or by similar or dissimilar methodology now known or hereafter developed. Exempted from this legal reservation are brief excerpts in connection with reviews or scholarly analysis or material supplied specifically for the purpose of being entered and executed on a computer system, for exclusive use by the purchaser of the work. Duplication of this publication or parts thereof is permitted only under the provisions of the Copyright Law of the Publisher's location, in its current version, and permission for use must always be obtained from Springer. Permissions for use may be obtained through RightsLink at the Copyright Clearance Center. Violations are liable to prosecution under the respective Copyright Law.

The use of general descriptive names, registered names, trademarks, service marks, etc. in this publication does not imply, even in the absence of a specific statement, that such names are exempt from the relevant protective laws and regulations and therefore free for general use.

While the advice and information in this book are believed to be true and accurate at the date of publication, neither the authors nor the editors nor the publisher can accept any legal responsibility for any errors or omissions that may be made. The publisher makes no warranty, express or implied, with respect to the material contained herein.

Printed on acid-free paper

Springer is part of Springer Science+Business Media (www.springer.com)

OVS gratefully dedicates this book to Daisaku Ikeda, Uday V. Singh, Indu Bala, and Late Prof. Ben M.J. Pereira in appreciation for their encouragement

Contents

1	**Introduction**		1
2	**Survival Mechanisms of Extremophiles**		9
	2.1	Survival and Potential Therapeutic Strategies	14
		2.1.1 Ectoine-Mediated Mechanism.	15
		2.1.2 Evolutionary Diversity.	16
		2.1.3 Increased Catalytic Activity	17
		2.1.4 Amino Acid Accumulation.	18
		2.1.5 Aggregation Resistance Strategies.	18
		2.1.6 Activation of the Nuclear Factor.	19
		2.1.7 Resistance to Cell Death	19
		2.1.8 Cellular Compartmentalization	20
		2.1.9 Overexpression of Heat Shock Protein Genes.	22
3	**Therapeutic Implications of Extremophiles**		25
	3.1	Radiation-Resistant Organisms	29
	3.2	Thermophiles	30
	3.3	Halophiles	31
	3.4	Acidophiles	32
	3.5	Mesophiles.	33
	3.6	Psychrophiles	33
	3.7	Geophiles	34
	3.8	Barophiles	34

4	**Challenges in Advancing Extremophiles for Therapeutic Applications**		37
	4.1	Isolation and Purification of Extremolytes	38
	4.2	Systems Biology of Extremophiles	39
	4.3	Extremophiles Like Other Organisms	40
5	**Conclusion**		43
References			45

Abstract

Extremophiles are a large group of organisms with the ability to thrive under extreme environmental conditions such as high and low temperatures, high salt levels, radiation, and high antibiotic concentrations. They have been the center of attention due to the remarkable benefits they may have for humanity. The extremophiles' survival mechanisms are being investigated in order to meet challenges associated with human health; understanding these mechanisms could result in solutions to various medical problems humans face. This article discusses their survival mechanisms and possible implications in medical processes, as well as the major issues and challenges involved in advancing the commercial exploitation of extremophiles.

Keywords Extremophiles · Microorganisms · Genes · Proteins · Metabolic processes · Medical applications · Challenges

Chapter 1
Introduction

Abstract Microorganisms that are able to thrive in extreme environment are known as extremophiles. The metabolites and products that are secreted through these extremophiles are of immense importance. The environments that these extremophiles thrive enable the study of the metabolites and carry them over with the hope that they will help in medical and therapeutic applications. As more information is processed about extremophiles, it is easier to identify them and classify them for their metabolic properties. Extremozymes, the enzymes that are secreted through extremophiles are used in the therapeutics of treating different medical conditions. Studies have showed that these enzymes are important in the therapies, yet a throughout understanding of these extremophiles will enable for more research to be done to prove that these products and primary and secondary metabolites will be of use in the medical field and help to treat diseases.

Keywords Extremophiles · Metabolic processes · Extremozymes · Therapeutics · Survival strategies

Extremophiles are a group of organisms that survive in extreme and harsh environments. Their identification, classification, and potential for commercial applications in medical and therapeutic fields are receiving increased interest and attention. Primary and secondary metabolic products of commercial significance, i.e., extremolytes and extremozymes, are among the substances that are currently being studied for their potential benefits as medical solutions to several human diseases (Copeland et al. 2013; Kumar and Singh 2013; Singh 2013; Chakravorty and Patra 2013). Thriving in harsh environments make extremophiles good candidates for the exploration of bioprocesses and biotechnological applications. An important characteristic of extremophiles is maintaining the stability of their existence in order to thrive in their environments. Identifying the essential genes responsible for the activity of the organisms, the stability of their chemical properties, and their adaptability under harsh environmental conditions may have medical and therapeutic applications (Margesin and Schinner 1994; Feller 2003; Furusho et al. 2005; Ma et al. 2010).

Exploration of extremophiles and their biodiversity in order to make them useful for developing processes and products to serve humanity begins with the categorization of these organisms (MacElroy 1974; Rothschild 2007). Studies have not even scratched the surface in identifying extremophiles from natural habitats: less than 1 % of the organisms have been identified and even fewer have been sequenced for their beneficial properties. Table 1.1 summarizes major known extremophiles and their living conditions. The identification of extremophiles thus far has already provided opportunities for industrial and medical use, as reviewed by several researchers (Adams 1993; Adams and Kelly 1998; Niehaus et al. 1999; Blumer-Schuette et al. 2012; Gabani and Singh 2013).

Within the past two decades, a number of studies have explored beneficial roles of extremophilic bacteria and their usage in medicinal practices (Corry et al. 2014; Gabani and Singh 2013; Ksouri et al. 2012). The exploration of extremophilic organisms in various ecological and biodiversity settings has increased through identification of the various environmental conditions under which the microorganisms live and thrive (Ksouri et al. 2012; Cavicchioli et al. 2011). The genome sequences of these extremophilic microorganisms have helped provide a better understanding of their potential roles in industrial and medical applications (Lin and Xu 2013; Majhi et al. 2013; Jaubert et al. 2013; Wemheuer et al. 2013; Shin et al. 2014). Considerable progress has been made in genome sequencing methods and processes for identifying by-products from extremophiles (Ksouri et al. 2012; Liu et al. 2013, 2014). Applications of extremophiles in processes, such as biocatalysis and biotransformation, show great promise for benefits to human welfare (Hough and Danson 1999; Singh 2013).

The application of beneficial characteristics of extremophiles for industrial and medical purposes is becoming more possible now that the processes by which they contribute such benefits are better understood than ever before. Table 1.2 introduces a few major therapeutic roles for extremophiles. Copeland et al. (2013) provide a schematic illustration to demonstrate one of these contributions. Figure 1.1 summarizes the conversion of starch and cellulose into glucose and fructose that can be fermented into ethanol with the use of extremozymes found in thermophiles. Understanding how extremozymes can be used and applied in medical and therapeutic processes requires a fuller understanding of the systems biology (i.e., genomics, proteomics, and metabolomics) of these microorganisms. The scientific community has recognized "Extremophiles" as an economically and ecologically important group of microorganisms (Baker-Austin and Dopson 2007; Liu et al. 2014).

Despite recent progress, practical applications for extremophiles are still in their infancy in many areas, including therapeutic and medical applications. A large number of potentially beneficial extremophiles have yet to be investigated, and development of large-scale cultivation processes for extremophile organisms is still in its early stages. In addition, further efforts are needed to improve our understanding of the stability of the substances derived from extremophiles (Fornbacke and Clarsund 2013). The progress made in scientific methods such as random mutagenesis, protein engineering, direct evolution, and DNA shuffling has shown

1 Introduction

Table 1.1 Types of extremophiles, major species, and their living extremities in the environment

Types of extremophile	Selective species	Living extremities	References
Radiation-resistant	*Deinococcus depolymerans*; *D. guangriensis*; *D. radiodurans*; *D. wulumuqiensis*; *D. xibeiensis*; *D. gobiensis*; *D. gradis*; *D. misasensis*; *Cellulosimicrobium cellulans* (UVP1); *Bacillus pumilus* (UVP4); *B. stratosphericus* (UVR3); *Enterobacter* sp. (UVP3); *Roultella planticola* (UVR1); *Aeromonas eucrenophila* (UVR4); *Arthrobacter mysorens* (UVR5a); *Micrococcus yunnanensis* (UV20HR); *Stenotrophomonas* sp. (YLP1); *Brevundimonas olei* (BR2)	Resistance to survive under ionizing radiation; UVR resistance >600 J m^{-2}	Gabani et al. (2012), Copeland et al. (2013), Gabani et al. (2014), Asker et al. (2011), Sun et al. (2009), Wang et al. (2009), Yun et al. (2009a, b), Yun and Lee (2009)
Thermophiles	*Geobacillus thermodenitrificans*; *Thermus aquaticus* (YT-1)	High temperature above 45 °C and up to 121 °C	Arena et al. (2009), Lin et al. (2011)
Halophiles	*Halobacterium* spp; *Haloferax* spp; *Haloarcula* spp; *Halomonas stenophilia* B-100	High salt concentration such as salter pond brines and natural salt lakes	Ma et al. (2010), Buommino et al. (2005), Oren (2002), Llamas et al. (2011)
Acidophiles	*Ferroplasma acidarmanus*	Extreme pH levels; pH level 0	Edwards et al. (2000), Baker-Ausin and Dopson (2007)
Alkaliphiles	*Alkaliphilus transvaalensis*	pH level 12.5	Takai et al. (2001), Kobayashi et al. (2007)
Mesophiles	Phyla included: proteobacteria, firmicutes, and actinobacteria	+11 to +45 °C Habitats include yogurt, cheese, and moderate environments	Zheng and Wu (2010), Widdel and Bak (1992)
Psychrophiles	Himalayan midge	−18 °C	D'Amico et al. (2006), De Maayer et al. (2014), Irgens et al. (1996)
	Polaromonas vacuolata	<15 °C	

(continued)

Table 1.1 (continued)

Types of extremophile	Selective species	Living extremities	References
Geophiles	*Geobacillus thermoglucosidasius*; *G. stearothermophilus*; *G. thermodenitrificans*; *G. thermopakistaniensis*	Habitats include rich soils as well as keratin-enriched environments	Espina et al. (2014), Lin et al. (2014), Siddiqui et al. (2014), Pennacchia et al. (2014), Yao et al. (2013)
Barophiles	*Moritella* spp; alphaproteobacterium	High hydrostatic pressure; 80 Mpa	Kato et al. (1998), Emiley et al. (2011)

Table 1.2 Influensive roles of extremozymes/extremolytes in therapeutics from major extremophiles

Type of extremophile	Influensive role of extremolytes/extremozymes in therapeutics	References
Radiation-resistant	Extremolytes, ecotines, and other natural compounds used to cope with the stress of different environments	Rastogi et al. (2010), Singh and Gabani (2011, 2013), Copeland et al. (2013)
	Decreases the risks of skin damage and skin- related cancers	
Thermophiles	"Thermozymes" showed increased resistance to denaturing chemical agents; "ability to crystalize"; stowed away the aggregate prone regions (ARPs)	Sælensminde et al. (2008), Cava et al. (2009), Irwin and Baird (2004), Thangakani and Kumar (2012)
	Prevent misfolding of proteins of Alzheimer's and Parkinson's diseases; certain genes found in thermophiles are used for the heat precipitation from the host cell is found to isolate and purify the expressed protein	
Halophiles Ex: Halomonas stenophila (B-100 and N-12)	Metabolism of carbohydrates; oxygenic and an oxygenic phototrophs, aerobic heterotrophs, fermenters, de-nitrifiers, sulfate reducers, and methanogens	Tomlinson et al. (1978), Oren (2002), Llamas et al. (2011), Ruiz-Ruiz et al. (2011), Molina et al. (2013)
	Prevent cancer, chronic inflammation, cardiovascular disorder, aging process; promote anticancer activity through synthetic lipids	
	Blocked the growth of human T-lymphocyte tumors	
	Exopolysaccharides (EPSs) as antitumoral agents	

(continued)

Table 1.2 (continued)

Type of extremophile	Influensive role of extremolytes/extremozymes in therapeutics	References
Acidophiles	Amylases, proteases, ligases, cellulases, xylanases, a-glucosidases, endoglucanases, and esterases; plasmids, rusticynin, and maltose-binding protein	Sharma et al. (2012)
	Used in evolutionary medicine; preventive measures for ulcer disease and gastric cancer; enzymes are used toward polymer degradation	
Mesophiles	Cellular components, including their membranes, energy generating systems, protein synthesis machinery, biodegradative enzymes, and the components responsible for nutrient uptake; biochemical substrate storage capability	Metpally and Reddy (2009), Insel et al. (2007)
	Used throughout in proteases, for developments and advancements in extremozymes/extremolytes	
Psychrophiles	Energy transduction; regulation of intracellular environment and metabolism; functioning of enzymes; and protein conformation; synthesis of the gene products are not prevented by cold shock in cold-adapted microorganisms; secrete special "antifreeze", also known as cryoprotectant molecules that aid in decreasing the water point within a cell	Margesin and Schinner (1993), Irwin and Baird (2004), Zecchinon et al. (2001)
	Membrane fluidity; aid in maintaining viral vaccinations stability; also provide less tension for a cell to maintain a certain temperature and atmosphere in order to survive; help cells to survive in harsh winters	
Geophiles	Enriched in proteins including keratin proteins, fungal growth diminishing with the aid of proteins	Weitzman and Summerbell (1995), Achterman and White (2011), Lakshmipathy and Kannabiran (2010), Arena et al. (2009), Barbara et al. (2013)
	Prevent stomach ulcers, as well as gastric cancers that can develop with bacterial infections	

(continued)

Table 1.2 (continued)

Type of extremophile	Influensive role of extremolytes/extremozymes in therapeutics	References
	from parasitic worms; prevent chronic inflammation; adjuvant agents in equilibrating the immune response in viral diseases	
Barophiles	Mechanotransduction; proteases; gene expression is regulated by the functions of barophiles; gene encoding RNA polymerase activity; gene *ompH* is found regulated in barophilic organism	Tan et al. (2006), Irwin and Baird (2004), Nakasone et al. (1996), Bartlett et al. (1993)
	Used in coping mechanisms for cells in order to avoid detrimental changes to the environmental pressure changes of the cell; ensure cells are viable in all situations	

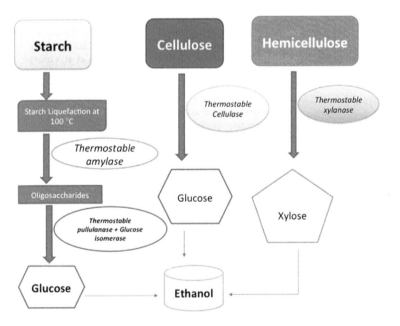

Fig. 1.1 Extremozymes and their production of sugars and ethanol from sources found in thermophiles (adopted and modified from Copeland et al. 2013)

promise for improving the stability and modifying the specificity of chemical substances derived from extremophiles (Danson and Hough 1998; Sellek and Chouduri 1999).

Given the breadth of this field, the current article aims to take a critical look at the survival strategies of the major extremophiles in order to investigate potential therapeutic targets for multiple disease types. There is a need to understand the molecular events in these organisms and their potential benefits in therapeutic applications. This article also brings together major issues and challenges in research areas related to extremophiles.

Chapter 2
Survival Mechanisms of Extremophiles

Abstract It is vital for extremophiles to cope with their environments making them viable to withstand under harsh environmental conditions. Extremophiles are known to adapt to the changes in their environment and surroundings that enable them to stabilize the changes in their homeostasis. The adaptability of extremophiles arrives from alteration of varying genes and proteins. Extremophiles produce extremolytes, which helps them to maintain their homeostasis such as ectoine-mediated mechanism, which is produced by halophiles and organisms alike. Evolutionary diversity, increased catalytic activity, amino acid accumulation, aggregation resistance strategies, resistance to cell death, activation of the nuclear factor, the use of heat shock proteins, and cellular compartmentalization, are all vital tools that extremophiles take on in order to conserve their genes.

Keywords Survival mechanisms · Proteins · Genes · Evolution · Diversity · Extremolytes · Metabolites

The general mechanisms that are studied and exploited in all the therapeutic and medical applications of extremophiles relate to how the extremophiles develop defensive mechanisms to survive in harsh environments and how their metabolisms are involved in these survival processes.

Extremophiles have been able to live in extreme and harsh conditions mainly due to their adaptability (Mallik and Kundu 2014; van Wolferen et al. 2013; Singh 2013). The adaptation mechanisms of such extremophiles would help researchers to understand their survival mechanisms, which in turn would help to figure out the process by which their molecular elements (i.e., proteins and genes) could be altered and used for therapeutic implications. Table 2.1 provides an overview of the survival and defensive strategies of selected extremophiles.

Singh and Gabani (2011) reviewed and described one such survival pathway in the radiation-resistant microorganism *Deinococcus radiodurans*. The microbial resistance against ionizing radiation induces pathway-specific genes, modulated proteins, and enzymes as part of the DNA repair mechanism. Figure 2.1 summarizes the survival strategy of *D. radiodurans*. This mechanism operates in three

Table 2.1 Survival and defensive strategies in major extremophiles to thrive under extreme environmental conditions

Extremophiles/extremolytes	Survival and defensive strategies	References
Thermophiles/carbohydrate extremolytes/hydroxyectoine	Stabilization of enzymes from stress and freeze drying; protection of oxidative protein damage; reduction of VLS in immunotoxin therapy	Kumar et al. (2010)
Halophiles/ecotines	Protection of skin immune cells from UV radiation; enzyme stabilization against heating, freezing, and drying; protection of the skin barrier against water loss and drying out; block of UVA-induced ceramide release in human keratinocytes	Buommino et al. (2005), Singh and Gabani (2011), Ortenberg et al. (2000)
Acidophiles/alkaliphiles	Maintaining a circumneutral intracellular pH; constant pumping of protons in and out of cytoplasm; acidic polymers of the cell membrane; passive regulation of the cytoplasmic pools of polyamines and low membrane permeability	Baker-Austin and Dopson (2007), Horikoshi (1999), Bordenstein (2008)
Psychrophiles	Translation of cold-evolved enzymes; increased flexibility in the portions of protein structure; presence of cold shock proteins and nucleic acid binding proteins; reduction in the packing of acyl chains in the cell membranes	Berger et al. (1996), Feller and Gerdey (2003), D'Amico et al. (2006), Chakravorty and Patra (2013)
Geophiles/EPS-V264; EPS-1,2,3	Mucoidal layer enveloping cell colonies; biofilm formation as stress response to extreme environmental conditions	Arena et al. (2009), Kambourova et al. (2009), Barbara et al. (2013)
Barophiles	Homeoviscous adaptation, tight packing of their lipid membranes; and increased levels of unsaturated fatty acids; polyunsaturated fatty acids maintain the membrane fluidity; robust DNA repair systems; highly conserved pressure regulated operons; presence of heat shock proteins	Lauro and Bartlett (2007), Yano et al. (1998), Rothschild and Mancinelli (2001), Kato et al. (1995 1996a, b), Kato and Bartlett (1997), Marteinsson et al. (1999)

distinct steps, as shown in Fig. 2.2. Three different pathways of survival have been identified through the process of homologous recombination, which is responsible for gene induction. First, the UVR-induced gene *uvrA* reveals uvrABC system protein A, representing a universal function in DNA repair and survival of

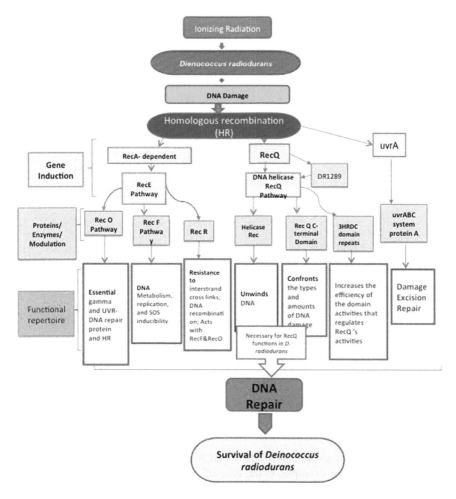

Fig. 2.1 A survival strategy of radiation-resistant microorganism *D. radiodurans* shows microbial resistance against ionizing radiation that induces pathway-specific genes, proteins, and enzymes of pathways in DNA repair mechanism (adopted with permission from Singh and Gabani 2011)

D. radiodurans (Fig. 2.2). This process of induction helps *D. radiodurans* to thrive in high-radiation conditions. UV induction in RecQ has been revealed to control DNA helicase, which further helps in managing the nature and the quantity of the DNA damage that is needed for RecQ functions in *D. radiodurans*. RecE pathway-dependent modulation in *recO* and *recF* reveals a functional repertoire of DNA repair protein, DNA metabolism, and replication and SOS inducibility. The modulation in *recR* could resist interstrand cross-links and DNA recombination acting with *recF* and *recO* (Singh and Gabani 2011). Thus, studying the defensive and survival mechanisms of the extremophiles in terms of their genome structure and the chemical properties of the compounds derived from them will help in the discovery of novel therapeutic and medical applications.

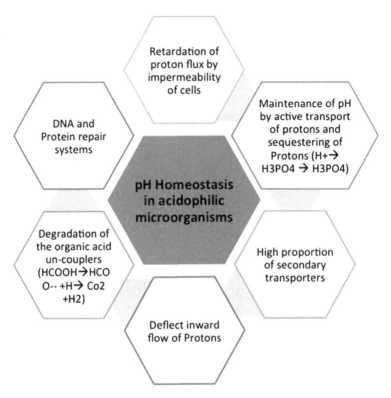

Fig. 2.2 pH homeostasis processes of acidophiles (adopted and modified from Baker-Austin and Dopson 2007)

Similar defensive mechanisms have been studied and described for other types of extremophiles. For example, in the case of acidophiles, which survive in highly acidic conditions, Baker-Austin and Dopson (2007) reviewed various survival pathways and mechanisms that enable these organisms to thrive at low pH. Impermeability of the cell membrane to protons is one such mechanism. Figure 2.2 summarizes various pH homeostasis mechanisms that have been identified (Booth 1985; Matin 1990). In general, the mechanism by which acidophiles use pH homeostasis has not been fully understood. However, efforts in sequencing the genomes of several acidophiles have shed light on several interrelated processes, including pH homeostatic mechanisms, impermeable cell membrane, cytoplasmic buffering, active proton extrusion, and organic acid degradation (Osorio et al. 2008; Cardenas et al. 2010; Liljeqvist et al. 2013; Guo et al. 2014).

Konings et al. (2002) describes the role of the cell membrane in the survival of bacteria and archaea found under extreme environmental conditions. The acidophiles have a rigid and impermeable cell membrane, which can restrict the cytoplasmic influx of protons. This helps to regulate the proton motive forces of the cell by determining the rate at which protons flow inward and pump outward

(Konings et al. 2002). Shimada et al. (2002) provided concrete evidence of this phenomenon in *Thermoplasma acidophilum*, whose cell membranes are made of tetraether lipids. Other examples of acidophiles include *Picrophilus oshimae* (van de Vossenberg et al. 1998a), *Sulfolobus solfataricus* (van de Vossenberg et al. 1998b), *Ferroplasma acidarmanus* (Macalady and Banfield 2003), and *Ferroplasma acidiphilum* (Golyshina et al. 2000; Batrakov et al. 2002; Pivovarova et al. 2002). Tyson et al. (2004) reported reconstruction of near-complete genomes of *Leptospirillum* group II and *Ferroplasma* type II and suggested that a wide variety of genes could be responsible for the impermeability of the cell membrane and preventing the inflow of protons to the cells. The above studies indicated that the genomes of organisms in microbial communities may reveal pathways for carbon fixation and nitrogen fixations, including energy generation, which will help us learn more about the survival strategies of microorganisms in extreme environments.

Michels and Bakker (1985) reported that bacteria such as *B. acidocaldarius* and *T. acidophilum* have exhibited the ability to actively pump protons out of their cytoplasm to maintain pH homeostasis. Such proton removal systems have also been reported in the *Ferroplasma* type II and *Leptospirillum* group II (*L. ferriphilum*) sequenced genomes (Tyson et al. 2004). Another key mechanism that acidophile cells use to maintain pH homeostasis is regulating the size and permeability of the cell membrane. Reducing the pore size of the cell membrane channels has been suggested as another mechanism to prevent protons from the acidic environment from entering the cell, thus helping maintain pH homeostasis. Amaro et al. (1991) characterized the outer membrane porin of the acidophile and revealed a large external loop that could be responsible for controlling the size of the pores in the cell as well as the ion selectivity. Guiliani and Jerez (2000) reported that at a pH level of 2.5, the external loop controlled the inflow of the protons across the outer membrane.

In the event of protons entering the cell membrane, acidophiles have a number of intracellular mechanisms to reduce damage that might be caused by the entering protons. The cells of acidophiles have a buffering mechanism to release the protons, as summarized in Fig. 2.2. This is possible because of the presence of certain cytoplasmic buffer molecules that contain basic amino acids such as lysine, histidine, and arginine that help in the proton sequestering process. Studying the cytoplasmic homeostasis of pH in the acidophilic bacterium *Theobacillus acidophilus*, Zychlinsky and Matin (1983) proposed that the amino acid side chains were primarily responsible for acidophile cytoplasmic buffering. Castenie-Cornet et al. (1999) supported this by finding that decarboxylation of amino acids such as arginine-induced cell buffering in *Escherichia coli* by consuming the protons and transporting them outside the cell membrane.

Another mechanism acidophiles use to maintain homeostasis is uncoupling the organic acids. This process is called cytoplasmic protonation, and is a result of the dissociation of protons in the cytoplasm. Researchers studying this organic acid degradation reported the authenticity of uncoupling reactions at low pH (Kishimoto et al. 1990; Alexander et al. 1987; Ciaramell et al. 2005).

Heliobacter pylori are known for causing gastric ulcers, and are able to survive in harsh acidic conditions. *E. coli* can survive in harsh acidic environments (pH 2–3) for shorter time spans, even though it prefers to be at neutral pH. The mechanisms that these two different microorganisms use to withstand acid in the stomach differ significantly. It is unclear how *E. coli* is able to survive the high acid levels in the stomach; however, studies have suggested three systems that enable microorganisms to resist high levels of acid for longer periods of time (Foster 2004). In the stationary phase, alternative sigma factor is what makes the cells tolerant to the various levels of acid. Another part of this mechanism is the cAMP receptor protein (CRP), which binds with the sigma factor to create a complex that tolerates high levels of acidity (pH 1–2) in the stomach. Acidophiles also have pumps that move protons in and out of the cell in order to neutralize the cytoplasmic membrane. This is required because when bacterial cells come into contact with extreme acid stress, as is the case with acidophiles, there is an influx of protons that decreases the internal pH of the cell (Foster 2004).

The defensive and survival mechanisms used by radiation-resistant and acidophilic organisms, as well as the other specific mechanisms that enable extremophiles to adapt to various environments, make them excellent candidates for exploring beneficial properties and therapeutic implications for multiple disease types (Furusho et al. 2005; Buommino et al. 2005; Kumar and Singh 2013; Copeland et al. 2013). However, the advantages the medical world can derive from these extremophiles are only in the early stages of recognition and realization. Some extremophiles may have the solutions, but the task at hand is to find what mechanisms can be effective in synthesizing potentially useful therapeutic products. To advance our therapeutic uses of extremophiles toward treatments of specific diseases in the future, it is necessary to have a better understanding of the physiology of these extremophiles.

2.1 Survival and Potential Therapeutic Strategies

It has been recognized that the characteristics that help extremophiles to survive in extreme environmental conditions could be effectively used in medical processes to develop applications that have benefits to human health. Radiation-resistant extremophiles have been reviewed to reveal their implications for developing anti-cancer drugs, antioxidants, and sunscreens (Singh and Gabani 2011; Gabani and Singh 2013). Similarly, thermophilic bacteria have been known to help in DNA processing, production of proteins and enzymes, and biotechnological processes (Oost 1996). Acidophilic bacteria contribute to acid mine drainage and help to neutralize the pH of certain cytoplasmic membranes by pumping protons into the cellular space (Edwards et al. 2000).

To advance the role of extremophiles in the search for specific therapeutic mechanisms and their implications, it is pertinent to ask what metabolic products such as extremolytes and extremozymes are produced and how these primary and secondary products can be effectively exploited for medical purposes. Here we discuss some therapeutic mechanisms of the selected extremolytes.

2.1.1 Ectoine-Mediated Mechanism

Aerobic, chemoheterotrophic, and halophilic organisms contain ectoine, which is chemically identified as (5)-2-methyl-1, 4, 5, 6-tetrahydropyridine-4-carboxylic acid. High levels of radiation can alter DNA structure and produce cancer unless the structure is repaired by cellular machinery. Copeland et al. (2013) reviewed and demonstrated usage of extremolytes in their setting, and proposed an ectoine-mediated hypothetical survival mechanism (Fig. 2.3). The mechanism of ectoine biosynthesis led to UV neutralization, revealing therapeutic implications of halophiles as summarized in Fig. 2.3. Halophilic extremophiles engage in a three-step process to produce ectoine from aspartate semialdehyde (ASA) (Fig. 2.3). Nakayama et al. (2000) reported that the gene cluster of *EctA*, *EctB*, and *EctC* encodes the enzymes needed for the synthesis of ectoine (Fig. 2.3A). Ectoine was produced through fermentation of *Halobacter elongate* in a continuous process and microfiltration of the biomass, and the ectoine filtrates were purified through electrodialysis, chromatography, and crystallization (Lentzen and Schwarz 2006). Skin is protected from UVA irradiation when human keratinocyte cells are pre-treated with ectoine (Bunger and Driller 2004). The ways in which ectoine may

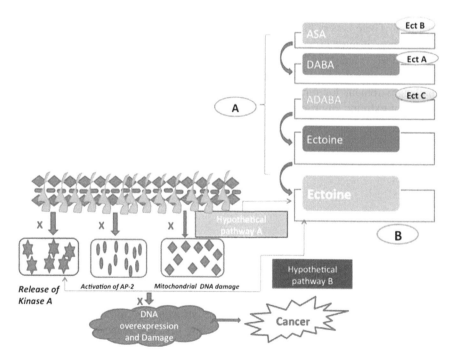

Fig. 2.3 Extremolytes in halophile bacterium *H. elongate* and proposed hypothetical survival mechanism. (*ASA* aspartate semialdehyde; *DABA* 1-2,4-diaminobutyrate; *ADABA* N-acetyl-1-2,4-diaminobutyrate) (adopted and modifed from Copeland et al. 2013)

help prevent damage to cells are shown in Fig. 2.3B: the release of secondary messengers (i.e., kinase), transcription factor AP-2 activation, intercellular adhesion molecule-1 expression, and mitochondrial DNA mutation. Beyer et al. (2000) demonstrated the immunoprotective effects of ectoine treatment through treating Langerhans cells under UV stress with 1 % ectoine. All these studies provide ample evidence of the protective properties of ectoine, which is hypothesized to provide protection from DNA damage and hence from cancer (Fig. 2.3B).

The therapeutic implication of this defensive mechanism is that ectoine helps stabilize the membrane structures, resulting in a higher level of resistance to UVA damage. Further, ectoine-mediated neutralization has been found to reduce or prevent dehydration of dry atopic skin and prevent skin aging (Singh and Gabani 2011). Similar roles of ectoine have been explored in research on apoptotic cell deaths in the contexts of Machado–Joseph disease (Furusho et al. 2005) and Alzheimer's disease (Kanapathipillai et al. 2005).

2.1.2 Evolutionary Diversity

Despite the diversity in living world, microorganisms are yet to see the tip of the iceberg. Most microorganisms existing in nature, particularly bacteria have yet to be identified. There is very little known to the current microbiologists on how to grow wide variety of microorganisms. In the sense of unknown growth medium for most microbial life, metagenomics have been considered to extract the total nucleic acid from environment with limited success to explore the hidden microbial life. The challenges remain for microbiologists to isolate novel microbial species from a variety of extreme environmental conditions.

Due to their biochemical properties, extremophiles are of high interest to both basic and applied microbiologists. Thermophiles contain DNA binding proteins, which have a potential role in maintaining DNA in a double-stranded form at high temperatures (Pereira and Reeve 1998). In order to diversify microbial community in the thermal environment, the heat-mediated alteration was reported to affect the membrane stability by opposing hydrophobic residues from each layer of the "lipid bilayer" membrane together forming the lipid monolayer instead of a bilayer that prevents the cell membrane to melt at high temperature (van de Vossenberg et al. 1998a). This diversifies the microbial survival at specific niche. However, at low temperature, proteins are being revealed to be more polar and less hydrophobic than proteins in thermophiles. In addition, psychrophiles regulate chemical composition of their membranes by maintaining the length and degree of unsaturation of fatty acids. This regulation keeps the membrane structure in sufficiently fluid form allowing transport process to occur, even below freezing temperatures (Horikoshi and Grant 1998).

Extremophiles have been reported to carry a set of essential genes that are evolutionarily conserved (Duplantis et al. 2010). These essential genes play an important role in translating the useful products that enable their survival under

2.1 Survival and Potential Therapeutic Strategies

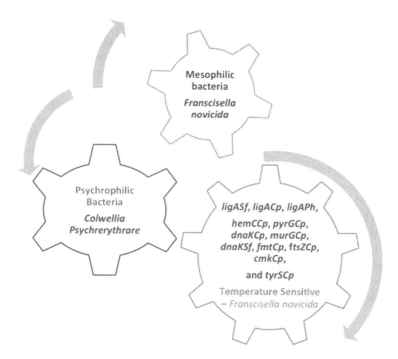

Fig. 2.4 Essential genes from a psychrophilic bacterium are transformed into temperature-sensitive mesophilic host organism (adopted and modified from Shanmugam and Parasuraman 2012)

harsh environmental conditions. The variations in extreme environmental characteristics exert pressure on essential genes (*ligASf, ligACp, ligAPh, hemCCp, pyrGCp, dnaKCp, murGCp, dnaKSf, fmtCp, ftsZCp, cmkCp,* and *tyrSCp*) that help extremophiles adapt to the environment. This phenomenon is referred to as "evolutionary diversity" and the properties of the essential genes could be used to engineer bacterial pathogens that are stable and temperature sensitive which further could be used as vaccines. Figure 2.4 summarizes the involvement of essential genes from psychrophilic bacteria transformed into temperature-sensitive mesophilic host organisms. Studies have also substituted essential genes of bacteria found in arctic environments for the genes in pathogenic organisms (Duplantis et al. 2010; Shanmugam and Parasuraman 2012).

2.1.3 Increased Catalytic Activity

Another extremophilic mechanism with the potential for therapeutic applications is the metabolic fluxes among psychrophilic microorganisms (Georlette et al. 2003). Psychrophilic organisms thrive in extreme cold habitats, produce enzymes that are active in cold temperatures, and can cope with the low-temperature-induced reduction in chemical reaction rates. The enzymes produced by psychrophilic organisms will have high catalytic efficiency at low temperatures.

It has been suggested that at the active sites, cold-adapted DNA ligase has specific characteristics such as high conformational flexibility, increased activity at low and moderate temperatures and overall destabilization of the molecular edifice (Georlette et al. 2003), revealing potential implications for biotechnology applications. These characteristics are reversed in mesophiles and thermophiles, which show reduced activity at low temperatures, high stability, and reduced flexibility. Because of the complexity involved in understanding these properties, large entropy changes are involved in the denaturation process of these microorganisms. The results of this study, conducted by adapting the different thermal habitats, indicated functional links between activity, flexibility, and stability. Studies have also been conducted on amylases and xylanases derived from extremophiles (Elleuche et al. 2011; Qin and Huang 2014; Liu et al. 2014). The therapeutic implications of increased catalytic activity in psychrophilic organisms due to the tradeoff between the low temperature and lower thermal energy resulting in specific changes in the molecular structure need to be further exploited.

2.1.4 Amino Acid Accumulation

Some extremophilic adaptation mechanisms that produce substances useful to humans are explained (Hendry 2006). While the enzymes produced by acidophiles and alkaliphiles can be useful in extreme conditions, the organisms themselves can also regulate their cytoplasmic activities at neutral pH conditions. Halophiles, however, adapt by regulating the salt concentration in their cytoplasm; the cytoplasmic proteins of the halophiles adapt to the environment by accumulating anionic amino acids on the cell surfaces. This property is also useful in improving their stability and activity in nonaqueous solvents. Halophiles also tend to reduce their osmotic pressure by gathering high levels of low-molecular-weight neutral organic species (Hendry 2006).

2.1.5 Aggregation Resistance Strategies

Maintenance of metabolic flux and cellular mechanisms relies upon the organisms' ability to keep their functional states when they are under extreme stress. By understanding the aggregation resistance strategies of thermophilic proteins, it is possible to resolve the response of the aggregation-prone regions in proteins. Thermophiles produce proteins that help in addressing the protein aggregation that reduces the functional state of the organisms (Merkley et al. 2011; Kufner and Lipps 2013). Thangakani et al. (2012) compared the aggregation resistance strategies adapted by thermophilic proteins and their mesophilic homologs using a dataset of 373 protein families and found that the thermophilic proteins had better utilization of the aggregation resistance strategies. Thermophiles tend to accumulate osmolyte molecules that can stabilize their proteins and macromolecules, which could help in the design and formulation of proteins and antibodies with therapeutic applications.

2.1.6 Activation of the Nuclear Factor

The ability of heat shock proteins (HSPs) to inhibit the genetic expression of proinflammatory cytokines has been explored as another mechanism by which extremophiles survive under harsh environmental conditions. Buommino et al. (2005) reported that the transcription of proinflammatory cytokines is dependent on the activation of the nuclear factor kappa-B (NF-kappaB). Studies indicate that ectoine, a biomolecule produced by halophiles, activates certain heat shock proteins. The authors used reverse transcriptase-polymerase chain reaction (RT-PCR) and immunoblot analysis to determine the increased levels of gene expression of HSPs in human keratinocytes that were treated with ectoine and heat stress. The findings had important implications for the development of additives that can be used as protective tools for treating human skin infections or inflammation.

2.1.7 Resistance to Cell Death

Other recent investigations have examined extremophiles' resistance to cell death and the pathways by which this process occurs, as shown in Fig. 2.5. One major hypothesis that has been supported by several studies involves the role of mitochondria in the death of brain cells: a set of protein components affects mitochondria and begins their destruction, leading to cell death under various conditions. Thus, further research on extremophiles and the proteins in their mitochondria may provide clues for identifying compounds that do not destroy mitochondria. Biochemical assays and protein sequencing will assist in identifying the mechanisms of molecular mediation in cell death. This can lead to the development of drugs to target proteins that cause cell degeneration and reduce the development of neurodegenerative diseases. This mechanism for using extremophiles in the development of therapeutic applications is summarized in Fig. 2.5.

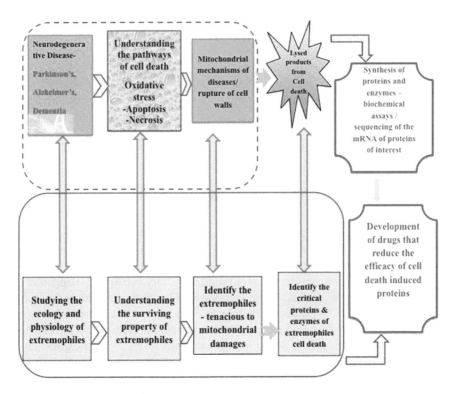

Fig. 2.5 A process of studying the mechanism for using extremophiles in the development of therapeutic applications for neurodegenerative diseases (based on: http://www.projectsmagazine. eu.com/randd_projects/mitochondrial_mechanisms_of_disease_lessons_from_extremophiles)

2.1.8 Cellular Compartmentalization

A UVR neutralization model using cellular compartmentalization of scytonemin biosynthesis in cyanobacteria was studied (Soule et al. 2009). Here, we attempt to summarize the possible therapeutic implications of this model (Fig. 2.6). The outer membrane of the cyanobacterium absorbs the UVA irradiation (Fig. 2.6, left), which further stimulates a cluster of genes such as *Tyrp*. This activates production of tryptophan and p-hydroxyphenyl pyruvate monomers from chorismate. In addition, it is proposed that certain precursors are processed by ScyA, ScyB, and ScyC and NpR1259 in the cytoplasm. Using these precursors, reduced forms of scytonemins are produced by perisplasmic enzymes (ScyD, ScyE, ScyF, DsbA, and TyrP). These reduced forms of scytonemin autooxidize from the extracellular slime layer in sufficient quantity to block the incoming UVR (Soule et al. 2009).

Singh and Gabani (2011) conceptualized this model for the eukaryotic cell. Scytonemin was anticipated to provide a novel pharmacophore for the development of protein kinase inhibitors as antiproliferative and antiinflammatory drugs. It was

2.1 Survival and Potential Therapeutic Strategies

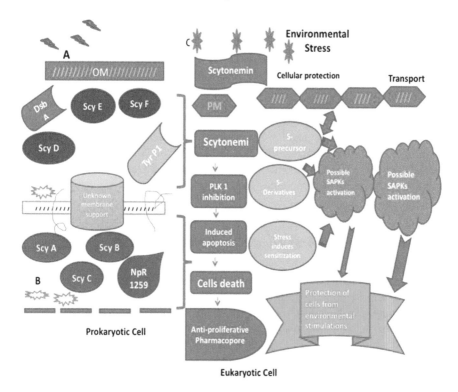

Fig. 2.6 Cellular protection of biosynthesized scytonemin in prokaryotes and hypothesized proposed mechanism of cellular protection in eukaryotic cell (adopted and modified with permission from Singh and Gabani 2011)

also hypothesized that scytonemin derivatives may be involved in the survival of healthy cells through mediated activation of stress-activated protein kinases (SAPKs), shown with dotted arrows on the right in Fig. 2.6.

Singh and Gabani (2011) reviewed the ATP-competitive inhibitors of polo-like kinases (PLKs), which have been theorized to control oncogenes in human cells, since they have the ability to switch off the activity by binding to ATP-binding sites. PLK1 has been highly regarded as a mitotic cancer target, and can be inhibited by scytonemin, which is recognized as a nonspecific ATP competitor. Further, scytonemin has the property to treat hyperproliferative disorders (Stevenson et al. 2002b). Luo et al. (2009) demonstrated that due to mitotic stress, cells become highly sensitive to PLK1 inhibition when they have mutant Ras acting as an oncogene. The scytonemin-mediated inhibition of PLK1 expression has been shown to induce apoptosis in osteosarcoma cells and other cancer cell types (Stevenson et al. 2002a; Duan et al. 2010).

These studies suggested that scytonemin could function as a novel pharmacophore for the development of protein kinase inhibitors and antiproliferative and antiinflammatory drugs (Fig. 2.6). In addition, SAPKs such as p38/RK/CSBP kinase and c-Jun N-terminal kinase (JNK) could help in the development of therapeutic responses to shock and UV-radiation-related stress. It is known that SAPKs when activated can further activate transcription factors (c-Jun, ATF2, and Elk-1) responsible for gene expression responses to external environmental stresses. Alternatively, Singh and Gabani (2011) anticipated that scytonemin-mediated activation of SAPKs could help in eukaryotic cell survival. There also exists some complementarity of the scytonemin activity responsible for the UV insensitivity of photosynthesis in *Nostoc flagelliforme* and the UV absorption of mycosporine-like amino acids (Ferroni et al. 2010). The therapeutic propositions of these studies indicate that there is complementarity between biologically mediated UV protection and the pharmaceutical compounds used for UV protection.

2.1.9 Overexpression of Heat Shock Protein Genes

HSPs have immunomodulatory properties that could have proinflammatory functions that help in immune responses. The mechanism that HSPs use to regulate autoimmunity can be effectively harnessed to develop therapeutic tools for the treatment of autoimmune disorders and some forms of cancer. Welch (1993) showed that high levels of HSPs could be achieved by exposing the cells to various types of chemical agents including metabolic poisons, heavy metals, protein modifiers, amino acid analogs, and ionophores. In another study, Zügel and Kaufmann (1999) demonstrated that during periods of stress caused by infection or inflammation, HSP synthesis could protect prokaryotic or eukaryotic cells. Boummino et al. (2005) reported that ectoine from halophiles may help in protecting cells from stress and prevent cell damage at higher levels of HSP70. HSPs are known for their role in the cytoprotection and repair of cells and tissues against the stresses and trauma they might face in extreme conditions (Morimoto and Santoro 1998).

HSPs involve overexpression of the single or collective HSP genes to help protect the skin from various stresses such as high levels of heat, drug toxicity, UV radiation, and other pollutants (Simon et al. 1995; Zhou et al. 1998). Among modulated variations in HSPs during cellular stress, HSP70 was revealed to be a major inducible and cytoprotective protein (Buommino et al. 2005). The overexpression of HSP70 significantly reduces the release of IL-6 induced by UVA, UVB irradiation, and oxidative stress (Buommino et al. 2005). Halophilic organisms were anticipated to be an effective source of small organic molecules that can be used to treat skin diseases that originate from infections or inflammation by overexpression of HSP70. Further, ectoine, a key extremozyme from the halophiles, could be effectively used as a protective additive for skin defense and in protein synthesis by increasing the basal levels of HSP70 (Buommino et al. 2005). Relatively similar mechanisms have been identified and documented in other extremophiles, and can be effectively applied in developing products that may have therapeutic values. Liu et al. (2010) reported that CiHsp70, a molecular chaperone of the HSP family, may

2.1 Survival and Potential Therapeutic Strategies

play a role in enabling Antarctic ice algae *Chlamydomonas* sp. ICE-L to acclimatize to the polar environment. Yamauchi et al. (2012) studied the protein-folding mechanism of the GroEL system in psychrophilic bacterium *Colwellia psychrerythraea* 34H, and found that the CpGroEL system has an energy-saving mechanism that allows it to avoid excess utilization of ATP to ensure microbial growth at low temperatures.

Chapter 3
Therapeutic Implications of Extremophiles

Abstract Extremophiles use specific mechanisms to alter the primary and/or secondary metabolites (i.e., extremolytes) to thrive under harsh environmental conditions, which could be exploited in therapeutics. Extremolytes from thermophiles, i.e., stable proteins, stable amylase, and thermozymes, have potential implications in regulation of intracellular environment and metabolism, and in energy transduction. Extremolytes mycosporine-like amino acids (MAAs), scytonemin, bacterioruberin, ectoine from radiation-resistant extremophiles can help to protect from UVR and gamma radiations. Extremolytes from acidophiles have been considered to use in protein pumps, reduce the pH of the cells within the cell surfaces, and as probiotics. Halophile carries unprecedented properties of carotenoids, high-antioxidant composition, and reductases. Further, investigations on extremolytes will pave the way toward next-generation medical innovation.

Keywords Extremolytes · Extremozymes · Extracellular polymeric substances · Neurodegenerative disease · Proteases · Amylase

The interest in the study of extremophiles stems from the fact that they are a good source of useful substances (i.e., extremolytes), including beneficial enzymes (i.e., extremozymes). These substances can be used as biocatalytic agents because of their unique characteristics, including their stability under extreme environmental conditions. Extremolytes are being applied in various ways in the medical field today. For example, they have been used in cancer research, in the form of anticancer drugs, antioxidants, cell-cycle blocking elements, and anticholesteric drugs. Extremozymes can be stable and active under conditions that are thought to be incompatible with normal biological organisms and materials (Hough and Danson 1999; Eichler 2001; Egorova and Antranikian 2005; Gabani and Singh 2013; Elleuche et al. 2014). One major potential use for extremozymes is the development of special industrial chemicals, agrochemicals, and intermediary products for the pharmaceutical industries (Demirjian et al. 2001). Table 3.1 summarizes the therapeutic implications of the substances that can be derived from extremophiles.

Table 3.1 Extremozymes found in extremophiles and their implicative products in therapeutics

Types of extremophile	Role in disease types	Extremolytes (proteins/enzymes) from extremophiles	Implications	References
Radiation-resistant	Prevents Alzheimer's and Parkinson's diseases from developing; Prevents damage from bone marrow and protects skin from exterior radiations; prevent skin damage due to from UVR; Skin cancer, etc.	Mycosporine like amino acids; scytonemin; bacterioruberin; ectoine	Extremolytes; extremozymes, used for their high stability; color pigmentation, help to protect from UVR and gamma radiations; increase of consumption of carbohydrates; anticancer drugs: UV absorption and sun screen	Copeland et al. (2013), Rastogi et al. (2012), Zhang et al. (2007), Singh and Gabani (2011), Yuan et al. (2009a, b), Karsten et al. (2009), Gabani and Singh (2013), Bunger and Driller (2004), Asgarani et al. (2000), Stevenson et al. (2002a, b)
Thermophiles	Treat misfolded proteins: aid in curing neurodegenerative diseases; thermolytes: used in protein degradation; suppress cancer activity: known to aid in T cell proliferation of leukocytes; antitumor suppression	Stable proteins; thermosomes, (chaperone protein), act as a facilitator and maintain the rest of the chaperone proteins for the cell; stable amylases; thermozymes are being produce in surrogate mesophilic hosts; enzymes and proteins from the genus Thermus, and thermozymes in general are good candidates for biocatalytical processes as they often present higher operational stability, and enhanced co-solvent compatibility	Phytosterols; cellulose production; energy transduction; regulation of intracellular environment and metabolism; functioning of enzymes; and protein conformation; citrate synthase from *P. furiosus*; and glucose dehydrogenase from *Thermoplasma acidophilum*	Jorge et al. (2011), Acharya and Chaudhary (2012), Margesin and Schinner (1994), Jorge et al. (2011), Kohama et al. (1994), Sellek and Chaudhuri (1999), Thangakani and Kumar (2012), Irwin and Baird (2004)
Halophiles	Prevent multiple cancer types; chronic inflammation; atherosclerosis; cardiovascular disorder; aging process;	Reductase from *H. volcanii* and malate dehydrogenase from *Haloarcula marismortui*; dihydrofolate reductase and	Motility capacity; unfolding and refolding of a halophilic enzyme has been studied; 3-hydroxy-3-methylglutaryl-	Ksouri et al. (2012), Agrawala and Goel (2002), Le Bail et al. (1998), Sellek and Chaudhuri

(continued)

3 Therapeutic Implications of Extremophiles 27

Table 3.1 (continued)

Types of extremophile	Role in disease types	Extremolytes (proteins/enzymes) from extremophiles	Implications	References
	anticarcinogenic flavonoids help to prevent cancer; promote antiproliferative cancer activity	dihydrolipoamide dehydrogenase; polyunsaturated fatty acids; carotenoids; vitamins; sterols; essential oils; polysaccharides; glycosides; phenyl compounds; high- antioxidant composition; proteins especially in halophiles require an excess of negatively charged amino acids on the protein surface; contains antifreeze proteins that can freeze body organs for study	coenzyme; inhibition of insulin and amyloid formation; reduction of apoptotic cell death; protection of mitochondrial DNA in human dermal fibroblasts	(1999), Buommino et al. (2005), Arora et al. (2004)
Acidophiles	Used in situ experiments; Fe cycling; Used in evolutionary medicine; preventive measures for ulcer disease and gastric cancer	Protein pumps are used excessively; peptidoglycan is used to reduce the pH of cells within the cell surface; biocatalysts; bio-enzymes, probiotics	Cultured media allows for special growth in amino acids; *Helicobacter pylori*: secretion of urease; enzymes stable at low pH such as amylases, proteases, ligases, cellulases, xylanases, α-glucosidases, endoglucanases, and esterases are known from various acidophilic microbes	Lu et al. (2010), Lopez De Saro et al. (2013), Sachs et al. (2011), Foster (2004), Irwin and Baird (2004), Sellek and Chaudhuri (1999), Sharma et al. (2012)
Mesophiles	Commonly used in proteases	Glutamate dehydrogenase; a protein that enables mesophiles to prevent the formation of protein aggregates. This enzyme is used for protein folding	Proteases; cold-adapted, enabling the microorganisms to adapt to changes in temperature	Fornbacke and Clarsund (2013), Singh et al. (2009), Sellek and Chaudhuri (1999), Thangakani et al. (2012)

(continued)

Table 3.1 (continued)

Types of extremophile	Role in disease types	Extremolytes (proteins/enzymes) from extremophiles	Implications	References
Psychrophiles	Membrane fluidity; frostbite; hypothermia; enzyme activation at high temperatures; marginal cellular growth	Flexible cellular proteins; antifreeze proteins; special proteins and molecular chaperones enable microbes to adapt and survive in cold temperatures; citrate synthase from an Antarctic bacterium, and α-amylase from *Alteromonas haloplanctis*; DNA polymerase from the uncultivated psychrophilic Archaeon *Crenarchaeum symbiosum*	Able to sense changes in temperatures: modulating membrane fluidity; leading bacteria to become temperature sensitive; aiding in establishing viral vaccinations; protecting against diseases that require cold-mediated immunity; increased enzyme production; lowering the temperature can change the flexibility of human proteins	Singh et al. (2009), Shanmugam et al. (2012), Margesin and Schinner (1994), Garcia-Descalzo et al. (2013), Selleck and Chaudhuri (1999), Gerday et al. (2000), Singh (2013), Irwin and Baird (2004)
Geophiles	Skin-related diseases; stomach ulcers; gastric cancers; help in finding treatments to prevent stomach worms as such	Protease could be an evolutionary adaptation in therapeutics; may lead to the decreased risk of derma-related diseases	Acid proteinases, elastase, keratinases; antibiotic cells; cell mediated immune reactions: can rid the fungus from the skin; leukocytes; antibodies	Weitzman and Summerbell (1995), Dahl (1987), Singh (2013)
Barophiles	Used in coping mechanisms for cells in order to avoid detrimental changes due to the environmental pressure; affect the cells working ability to adapt in different environments (pressures)	P60 protein was discovered to have pressures below standard condition; analysis was done on the protein and found heat shock properties in the protein and resembling archaic organisms; stress proteins are mostly expressed in barophilic organisms	Changes in membrane fluidity; amounts of fatty acids; changes in biochemical processes; Pressure-sensing mechanisms/organs: pressure-regulated operons	Marteinsson et al. (1999), Yano et al. (1998), Kato and Bartlett (1997), Marquis and Keller (1975)

3.1 Radiation-Resistant Organisms

Radiation-resistant organisms have been studied for the development of medical applications that can help humans suffering from radiation consequences such as skin cancer and premature aging. Mycosporine-like amino acids (MAAs) are found in diverse extremophiles, including cyanobacteria and eukaryotic algae. MAAs absorbs radiation with wavelengths of 310–365 nm, and are commercially used in sunscreens and other cosmetic purposes.

MAAs are known to prevent the formation of DNA dimers, which damage DNA when cells are exposed to UVR. MAAs have also been shown to protect fibroblast cells in human skin from UVR exposure (Oyamada et al. 2008). MAAs such as palythine, asterina, palythinol, and palythene have been suggested as prime candidates for UV protection. Some of these MAAs are found in several species of microalgae such as *G. galatheanum* and *G. venificum* (Llewellyn and Airs 2010). Torres et al. (2006) reported the isolation of mycosporine with UVB-absorbing properties from *Collemacristatum*, a lichenized ascomycete. In cultured human keratinocytes, MAAs exhibited protection from UVB-induced processes, such as erythema, pyrimidine dimer formation, and other types of membrane destruction. Researchers have suggested the use of MAAs to prevent UVR-induced cancers such as melanoma (De la Coba et al. 2009).

Mukherjee et al. (2006) suggested that maristentorin found in *Maristerntor dinoferus* could have properties to protect from UV irradiation. In addition to MAAs, bacteriorubrins have also been shown to be highly resistant to ionizing radiation. Bacteria such as *Rubrobacter radiotoleransis*, *D. radiodurans*, and *Halobacterium salinarium* have been reported to have considerable resistance to UVR and H_2O_2 (Asgarnai et al. 2000). In a study using a halobacterium that contained bacterioruberin, Shahmohammadi et al. (1998) found that the presence of bacterioruberin was also found to have DNA repair mechanisms.

Sphaerophorin and pannarin derived from lichens have been shown to have antioxidant activity, which is useful in sunscreen products for UV protection (Muller 2001). Russo et al. (2008) evaluated the effect of sphaerophorin and pannarin on chemically induced DNA cleavage and found that both compounds showed a protective effect on plasmid DNA. These compounds inhibited the growth of human melanoma cells (M14 cell line) by inducing apoptosis in cell culture experiments (Russo et al. 2008).

Studies indicate that substances from extremophiles such as bacterioruberin, sphaerophorin, and pannarin play a significant role in the development of potential treatments for diseases caused by excessive radiation, as reviewed by Singh and Gabani (2011) and Copeland et al. (2013). The therapeutic implications of these compounds for repairing DNA damage and thereby treating the diseases are yet to be explored.

3.2 Thermophiles

Thermophiles require high temperatures to thrive, and have several medical applications because their cell membranes have evolved to endure intense heat. As a result, these extremophiles are in demand for treating types of cancers involving intense heat. Since global warming has and will continue to have an impact on the environmental temperature, there has to be ways to reduce the effects of high temperature on human lives. Thermophiles have a mechanism in which they are able to convert carbohydrate-rich substances into hydrogen, and because they are able to withstand and grow at extremely high temperatures, this makes them applicable for use in the medicinal and therapeutic formulations. One major advantage that thermophiles may have is that they prevent the growth of pathogenic organisms from entering into the human society, since they survive at high temperatures at which most pathogens would not be able to proliferate (Suzuki et al. 2013).

Major neurodegenerative diseases such as Huntington's and Parkinson's are characterized by protein folding malfunctions and the inclusion of formations inside neurons in the brain (Jorge et al. 2011). Thermophiles may have extremozymes with capabilities such as the inhibition of conformational changes of native or mutant proteins, which may be useful to develop cures for neurodegenerative diseases (Jorge et al. 2011). Jorge et al. (2011) suggested that novel organic solutes known as thermolytes could serve to protect proteins' native structures. Thermolytes have been revealed to affect the growth curves of HEK293 cells, which have been cultured to aid in the treatment of Parkinson's disease. The results of this research showed a decrease in cell density in the cells that were treated with thermolytes compared to untreated control cells (Jorge et al. 2011). Studies were conducted to test the similar effects of thermolytes in Huntington's disease using transfected HEK293 cells with Htt-103QEGFP, and revealed an Mg- and DGP-mediated significant effect on the formation of Huntington aggregates (Jorge et al. 2011).

Identifying the missing components in cytoplasmic disulfide bond formation in hyperthermophiles where protein folding occurs may provide understanding of this new property that may lead to the discovery of antiviral drugs and large-scale production of therapeutics (Saaranen and Ruddock 2013). The interesting part of this study is the role of the cytoplasmic machinery of specific extremophiles, as that is where the protein folding takes place, and it is an important part of antiviral drug production (Saaranen and Ruddock 2013). Another emerging global challenge that is a result of increased temperatures is the scarcity of water, because the heat evaporates the water in reservoirs. This has become an increasing challenge for the water deprived communities of the world including in the North American state of California. Conditions that help mesophilic organisms to survive at high temperatures and water deprived environments need to be understood further to develop outcomes that may be useful to keep animals and humans less dependent on water.

3.3 Halophiles

Halophile archaea and bacteria are found in environments with high salt concentrations. The extremolytes from halophiles are categorized by their low-molecular mass and accumulation to stress, usually in response to salt and temperature (Chakravorty and Patra 2013). However, a major setback in using these extremolytes is that they have not been cultured in cells with higher cellular density. The biodiversity of the halophiles has been better understood and their therapeutic implications have improved due to knowledge of the role that halophilic organisms inhabit in the saltern crystallizer ponds (Oren 2002). Improved availability of chemotaxonomic studies and the effective use of advanced culture techniques and molecular biological methods have led to more effective exploitation of halophilic microorganisms.

Lipids derived from halophilic bacteria have found uses as deliverers of drugs and vaccines (Abrevaya 2013). Other compounds found in halophilic archaea, such as siderophores, offer iron-chelating agents that can be used to treat iron deficiency diseases or to increase antibiotic activity against bacteria (Abrevaya 2013). Oren (2002) reported that halophiles' metabolic activities decrease in diversity when the level of salinity they live in decreases. The level of energy generated by halophiles and the energetic cost involved in their osmotic adaptation are increasingly directly connected to high levels of salinity.

In general, extremophiles have a special characteristic in that they contain extracellular polymeric substances (EPS). Although the therapeutic mechanisms involved in these substances are not yet fully understood, they are known to help extremophiles and other prokaryotic organisms adapt to changes in their environments and compensate for the deleterious effects of harsh environments (Barbara et al. 2013). This is true for thermophiles and halophiles, as well as mesophiles that may have high industrial value.

Neurodegenerative diseases such as Machado-Joseph disease are categorized by mass protein misfolding as a key event in the pathogenesis. The overexpression of chaperone proteins recognizes misfolding as a typical target for effective therapy. Furusho et al. (2005) explored molecules from extremophiles that can potentially influence protein folding. Ectoine, which was originally found in halophiles, is an organic molecule of low-molecular mass and serves as an osmoprotectant; it is good at preserving enzymatic activity against freeze-thawing treatments of protein stabilization (Furusho et al. 2005). Ectoine was observed to reduce large cytoplasmic inclusions and increase the frequency of nuclear inclusions, although the integrity of the nuclei appeared to be maintained. Further, ectoine was shown to protect cells from polyglutamine-induced toxicity (Furusho et al. 2005).

Ectoines are also helpful in treating Alzheimer's disease. They play an important role in inhibiting the formation of amyloid, which is a protein aggregation factor involved in the misfolding of proteins (Kumar and Singh 2013). A majority of extremolytes comes from marine organisms, including mannosylglycerate (firoin) and mannoslyglyceramide (firoin A). Found abundantly in thermophilic bacterium

Rhodothermus marinus, these extremolytes are used against certain cancers and related diseases (Kumar and Singh 2013).

Buommino et al. (2005) showed that by the induction of HSP70 protein at elevated levels, ectoine exhibited a cytoprotection effect through bacterial lipopolysaccharide. Overexpressions in HSP70 and HSP70B9 were observed in keratinocyte cells treated with ectoine along with the heat shocks. However, this study was based on earlier findings, which showed that exposure to chemical inducers lead to both HSP induction and inhibition of NF-kB (Thanoas and Maniatis 1995). NF-kB is a eukaryotic transcription factor that can be induced by bacterial and viral infections, inflammation, and UV radiation. Its activation is a result of phosphorylation and degradation of the inhibitory protein IkB-a. Buommino et al. (2005) evaluated the degradation of IkB-a and showed that ectoine did not activate NF-kB in treated keratinocyte cells. Yoo et al. (2000) demonstrated that the induction of HSPs inhibits proinflammatory cytokine expressions. Further, the same studies indicated that the induction of HSPs can block the nuclear translocation of NF-kB by inhibiting the degradation of IkB-a. Thus, it can be interpreted that keratinocytes exhibiting cytoprotection mechanism could be a result of the ability of ectoine to induce HSPs and downregulate proinflammatory signals. The therapeutic implication could possibly be to prevent water loss in dry skin and aging of the skin through ectoine-mediated neutralization of UVR.

3.4 Acidophiles

Acidophiles are widely used for therapeutic purposes, and have even been involved in the evolution of medicine. They are mostly used in preventing gastric cancers and stomach ulcers for those prone to infections (Foster 2004). However, they have also been found useful for iron cycling and conducting in situ experiments. The acidophilus organisms are collectively known as probiotics and intestinal inhabitants. Some of these internal inhabitants are more helpful than others. Acidophilus aids digestive tract function and reduces the presence of harmful organisms. For this reason, use of probiotics can help prevent infectious diarrhea. *Lactobacillus acidophilus* has been used to prevent many diarrhea infections, such as traveler's diarrhea, infectious diarrhea, and antibiotic-related diarrhea (Elmer 2001; Lievin-Le Moal 2007; Grandy et al. 2010; De Vrese et al. 2011; Ouwehand et al. 2014).

Crohn's disease and ulcerative colitis conditions are known as inflammatory bowel diseases. Chronic diarrhea is a common feature of both conditions. Microorganism *Helicobacter pylori* is known to cause ulcers in the stomach and duodenum. Studies have been performed that used probiotics to inhibit the growth of *H. pylori* (Michetti et al. 1999; Pantoflickova et al. 2003; Gotteland et al. 2008). A thorough review by Elmer (2001) on microorganisms *Lactobacillus*, *S. boulardii*, and other probiotics revealed an in-depth understanding of the use of acidophiles for varying forms of diarrhea, and presented them as helpful for mild diarrhea in stable Crohn's disease.

3.4 Acidophiles

The acid stability of some acidophiles has been studied through analysis of their crystal structures. Acidophiles produce some substances that may be used in novel drug treatments of those prone to infections from gastric cancers and stomach ulcers (Lopez de Saro et al. 2013). To prevent fluctuations in the cellular membrane, these organisms pump hydrogen ions into the membrane, which regulates the pH of the cell (Irwin and Baird 2004). Nonetheless, exactly how acidophiles use these pumps to maintain the pH and stability of their cells is still not understood (Irwin and Baird 2004). Bioenzymes have been found in the genomes of acidophiles, and researchers are interested in cultivating them for use in biotechnology (Lu et al. 2010).

3.5 Mesophiles

Mesophiles are known for their ability to thrive at moderate temperatures and pressures. These extremophiles are readily able to compose themselves in a way that enables them to cope with the different surroundings and environments that they have to endure. One major example of products that allow mesophiles to survive is proteases. Proteases are enzymes that are used everyday in medical fields (Fornbacke and Clarsund 2013) and can adapt to cold temperature changes easily (Singh et al. 2009). They can also be used in combination with different extremophiles, such as psychrophiles, geophiles, and so on. Medical technicians are now able to use the properties of mesophiles toward isolating, creating, and supplying vaccines, which need to be kept at viable temperatures so that they do not start to lyse. In other words, cold shock for mesophiles is an important part of the cell cycle (Piette et al. 2012).

3.6 Psychrophiles

Psychrophiles adapt quickly to cold temperatures. These microorganisms have immense potential for the medical field, as they can provide insights into how cells cope with cold environments as well as actively protect and regulate the membrane fluidity of the cell. The cell membrane is the most important factor when it comes to psychrophiles, as it controls and regulates the homeostasis of the cell. If this equilibrium is offset, it causes damage to the cell in the long run (Margesin and Schinner 1994). It is important for psychrophiles to maintain certain temperatures and conditions to manage changes in external cellular lipid saturation and the disruption of intracellular organization (Margesin and Schinner 1994). Psychrophiles are commonly used in protecting against diseases that require cold-mediated immunity.

Polyunsaturated fatty acids produced from psychrophiles can be extensively used in pharmaceutical agents. Proteases have been used in medicine for many years in the treatment of blood disorders; they also have promising indications for treating digestion problems (pancrelipase) and muscle spasms, and as cosmeceuticals. Proteases are an established and well-tolerated class of therapeutic agents (Craik et al. 2011). Cold-adapted proteases have been used in a wide range of applications, including molecular biology, cosmetics, and pharmaceuticals (Craik et al. 2011; Gudmundsdottir and Palsdottir 2005; Marx et al. 2007). Cold-adapted proteases have been reported to be particularly useful in low-water conditions and high level of structural rigidity (Karan et al. 2012). Psychrophilic proteases have been obtained from Atlantic cod (*Gadus morhua*) and Antarctic krill (*Euphausia superba*). A wide variety of proteases have already been identified and genetically expressed in microorganisms (Taguchi et al. 1998), and cold-adapted proteases have been reviewed as an emerging class of potential therapeutics (Fornbacke and Clarsund 2013). Certain psychrophiles have been studied for their potential to prevent fungal infections (Garcia-Descalzo et al. 2013).

3.7 Geophiles

Geophilic or "soil-loving" microorganisms are a special type of microbial species; they live in soil and cannot be reproduced in laboratory settings. Due to this fact, geophiles can be useful in research that requires organisms in their natural settings. Geophiles are known for producing treatments for skin damage, stomach ulcers, and gastric cancers. They have also been used to help find preventive techniques/measures for decreasing the risk of infection (Dahl 1987). Products that are involved with the mechanisms of geophiles are acid proteinases, keratinases to help provide keratin, and antibiotic cells to prevent damage (Weitzman and Summerbell 1995). Arena et al. (2009) suggested that the biofilm formation of *Geobacillus thermodenitificans* acts as an adjuvant agent in equilibrating immune response in viral diseases.

Dermatophytes are a leading cause of fungal infections today (Achterman and White 2011). Costs for treating these fungal infections can be high; geophiles may make it possible to create an enduring treatment and cure for these types of fungal infections. Understanding the virulence of the fungi involved is important for coming up with the correct type of treatment, which is where the application of geophiles comes into play (Achterman and White 2011). Geophiles now have great promise for upcoming research in the field of therapeutics.

3.8 Barophiles

Barophiles survive at high-pressure levels, and can be found in environments such as deep-sea vents, high mountain ranges, and places where there is less oxygen (Kumar and Singh 2013). Barophiles are important in medicine because they can thrive at high pressure and are not affected by frequent changes in pressure levels.

3.8 Barophiles

Barophilic cells are most vulnerable to the aftereffects of pressure changes because it is important for biological cells to have stable pressure. Pressure changes can have detrimental effects that lead to cellular consequences (Tan et al. 2006). Barophilic microorganisms use mechanotransduction, a mechanism that senses the changes in pressure and translates them to a signal that can be used by the cell efficiently (Tan et al. 2006). Unwanted stress due to moderating pressures from different environments can cause this to occur (Tan et al. 2006).

It has been known for some time that pressure is involved in treating diseases; pressure affects all living organisms, and the right range of pressure is necessary for cell stability (Kato and Bartlett 1997). Barophiles survive under high-pressure ranges and can adapt to changes in pressure as well as changes in biochemical processes (Kato and Bartlett 1997). Barophiles can play an important role in therapeutics because of unusual characteristics such as abundance of pressure-sensing mechanisms on their cellular membranes, osmotolerance, and pressure-regulated operons (Marquis and Keller 1975). Tan et al. (2006) reviewed how human and prokaryotic cells respond to mechanical forces in order to identify how eye cells respond to pressure-induced glaucoma. Barophiles' ability to survive under extreme pressure may provide insights to develop treatments for pressure-induced injuries such as concussions and related athletic injuries.

Chapter 4
Challenges in Advancing Extremophiles for Therapeutic Applications

Abstract Varying types of extremophiles have been identified, however, many more are yet to be discovered from rare earth habitats. The challenges remain to grow these bacteria in the laboratory without knowing the optimum growth media and conditions. Advancements made in molecular biology of extremophiles are too limited to investigate the routes extremophiles adopt for themselves at molecular level under harsh environmental conditions. However, modern biology such as the "–omics" making it easier for researchers to sequence entire genome of bacteria and explore the systems biology approaches that enables extremophiles to cope with its surroundings. Reference libraries for chemical properties of extremolytes would make it easier to screen a variety of extremolytes against specific diseases. Developing bioreactors for efficient production of extremolytes is among the major challenges toward commerical benefits of extremophiles.

Keywords Isolation · "-omics" · Systems biology · Genomics · Proteomics · Metabolomics

Extremophiles have the potential to bring together multitudes of actors in the health industry and in the greater field of health treatments. Recent advances in the study of extremophiles show great promise for these organisms in therapeutic and medical applications. However, several issues and challenges stand in the way of developing useful medical applications of extremophiles.

First, the process of cultivating extremophiles in laboratory settings has been extremely cumbersome. The slow growth of the organisms, low yield of the required substances, and specialized equipment needed to cultivate them have limited the production of these organisms. Second, although advances have been made in the field of gene transfer, there are still no suitable gene transfer mechanisms for some extremophile groups such as *Archaea*. Third, a common host for enabling gene expression has not yet been found. Fourth, the review above indicates that the interest of researchers has been highly oriented toward studying enzymes with structural integrity that lend themselves to medical applications. Finally, in order to use extremophiles in therapeutic applications, various species must be studied in the extreme environmental conditions where they live. Such environments present harsh working conditions (i.e., radiation, temperature, etc.) for scientists, however beneficial the extremophiles may be (Mantelli 2003; Asker et al. 2011).

Irwin and Baird (2004) reported that protein function and structure are important indicators of potential therapeutic applications. However, further research is needed to explore the properties and uses of the enzymes derived from the proteins and their corresponding extremophiles. Advances made in this area, although limited so far, indicate that looking at the ways extremophiles work on a molecular level is the first step toward finding cures for specific diseases. The research reviewed above indicates that there has not been much research into products other than extremolytes and extremozymes that might be useful for therapeutic reasons. The families within these two types of products, such as ectoines and bioenzymes, show that extremozymes have a multitude of possible uses in curing diseases as well as providing insights and building connections between the medical world and the extremophilic bacterial kingdom.

4.1 Isolation and Purification of Extremolytes

The limitations involved in the study of extremophiles can be observed in fields that deal with particular types of extremophiles and the uses for which they are being explored. One of the most challenging factors in the development of medical uses for extremophiles is the difficulty of isolating and purifying extremolytes. Even when appropriate extremolytes are isolated and their uses identified, their development into specific drugs is a long and tedious process. Further, current regulatory procedures hinder speedy development of these drugs. This is mainly due to the challenges of limited functional analysis of specific molecules from extremophiles.

In the case of radio-resistant microbes, much research needs to be done before any significant contribution can be made toward development of drugs. Singh and Gabani (2011) reported that although radiation-resistant microbes contain compounds that could be harnessed to produce radioprotective drugs, research efforts remain insufficient. They suggested several reasons. First, unknown applications of UVR reservoirs could have marginalized the market requirement for a strategic product. Second, since the growth conditions and nutritional requirements of the extremophiles are highly specific, isolation, and maintenance of radiation-resistant microbes remains a challenge. Third, the instability in terms of their genomes, possible mutation, and pathogenesis increases the risk for scientists cultivating them in laboratory conditions. Finally, the limitations in terms of extraction and purification of the enzymes place constraints on their production process.

4.2 Systems Biology of Extremophiles

The systems biology-based high-throughput experimental approach can analyze global components from biological system (Ishii et al. 2007). These global components may help predicting the systematic behavior of cellular development and adaptation under extreme environmental conditions. This integrative approach of systematic behavior allows researchers to study the complex metabolic and regulatory networks of extremophiles in the microbial system, which may lead to the design of new value-added products of therapeutic significance. Now that 26,716 bacterial and 421 archaic genomes have been sequenced and abundant information is available at NCBI (http://www.ncbi.nlm.nih.gov/genome/browse/), the next challenge is to determine the function of each gene. Genome sequences of many extremophiles revealed unique genetic elements of potential significance (Tyson et al. 2004; Osorio et al. 2008; Lu et al. 2010; Cardenas et al. 2010; Lin and Xu 2013; Majhi et al. 2013; Jaubert et al. 2013; Wemheuer et al. 2013; Liljeqvist et al. 2013; Guo et al. 2014; Shin et al. 2014). Understanding of genetic roles in metabolic and regulatory networks lies ahead to grasping the functionality of biological system.

Functional "-omics" consists of high-throughput global experimental approaches that make use of the information and reagents provided by structural genomics to assess gene function (Hieter and Boguski 1997). This field has seen considerable growth in recent years, encompassing areas such as transcriptomics (global gene expression, i.e., mRNAs), proteomics (global proteins expression), and metabolomics (global expression of primary and secondary metabolites). Comprehensive transcriptome information for the extremophile Arabidopsis relative *Thellungiella salsuginea* provides firsthand clues of functional genomics elements in plant stress tolerance (Lee et al. 2013). A number of proteomics studies on a variety of extremophiles including *Acidithiobacillus ferrooxidans* (Chi et al. 2007; Osorio et al. 2013; Almarcegui et al. 2014); *Pyrococcus furiosus* (Lee et al. 2009); *Acidithiobacillus caldus* (Mangold et al. 2011); *Exiguobacterium* sp. (Belfiore et al. 2013); *Sulfolobus solfataricus* (Kort et al. 2013); *Methylacidiphilum infernorum* (Jamil et al. 2014); *Metallosphaera cuprina* (Jiang et al. 2014) reveal multiprotein-mediated exertion in the survival mechanism of extremophiles under a variety of environmental conditions. A global metabolomics study of extremophile is yet to come. In the mean time, the development of scientific methods for identifying, screening, and detecting microbial metabolites using advanced genomics, proteomics, and metabolomics methods show increased promise for understanding the structural and biochemical properties of extremophiles and extremozymes (Singh 2006; Karsten et al. 2009). Chemical reference libraries will help speedup progress in screening newly identified extremozymes for their therapeutic potential.

Hendry (2006) predicted that advances in the use of extremophiles for medical and therapeutic applications would require the use of novel methods to identify new species of extremophiles and innovative ways to develop and employ extremozymes. Mapping the genomic information of these useful organisms is still in its infancy. However, since the financial cost of genome and proteome sequencing and the time involved in the process have dramatically decreased in recent years, there is a high probability that the broader categories of extremophiles will be identified and their genomes and proteomes will be sequenced. Further, the altered products of primary and secondary metabolites (i.e., the metabolome) in biochemical pathways could be tracked using traditional techniques (i.e., NMR, GC-MS, LC-MS, etc.). This will further help in sequencing a full gamut of extremophiles, leading to a new level of understanding of the nature and properties of extremophiles and their use in medical applications.

4.3 Extremophiles Like Other Organisms

Given that the applied research on extremophiles has focused on a select few organisms, such as *E. coli* and *Helicobacter pylori* in the case of the acidophiles, exploration of the use of other species to translate the beneficial characteristics into therapeutic applications remains a challenge (Baker-Austin and Dopson 2007; Woappi et al. 2014). Further, due to recent developments in genome sequencing techniques, much of the research has been on developing hypotheses related to the

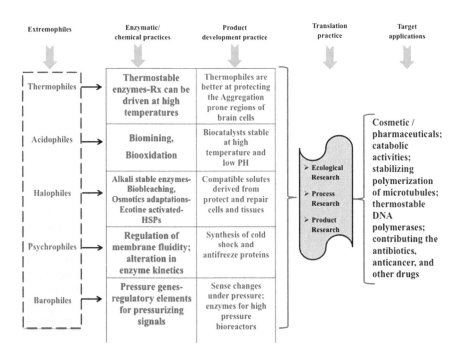

Fig. 4.1 Extremophiles, representative chemical processes, and product development mechanisms for medical applications

genomics data and little has been done to test these hypotheses. Baker-Austin and Dopson (2007) highlighted the need for genetic tools to perform in-depth analysis of the genetic and biochemical foundations of the observed phenomena and defensive mechanisms. Another technological challenge in developing extremophiles is the specific peculiarities of these organisms, which are extremely difficult to replicate in a natural environment. Further, a lack of genetic elements for vector development in the genetic markers continues to pose serious challenges to the development of therapeutic applications for compounds from extremophiles. However, the challenge remains to move forward with serious applications of extremophiles that will require innovative methods of prospecting these organisms from their natural habitats, as well as developing genetic and biotechnological approaches to understand the molecular and structural mechanisms and the biochemical strategies that these organisms employ to survive in harsh and extreme environmental conditions (Fig. 4.1).

Chapter 5
Conclusion

Abstract Even with the amount of limitations that have come across in this research and novel area of interest, a great amount of information has been found on the coping mechanisms and thriving skills that extremophiles use in order to live in their environments. With more interest coming to this area, it will allow to broaden its horizons developing tools to help replicate the environments in which extremophiles live. Collaborative efforts exploring extremophiles would ultimately be of benefit to human society.

Keywords Extremolytes · Extremozymes · Human society · Pharmaceutical industries · Sustainable therapeutics

Due to the complex properties of extremophiles and extremolytes, research in this field will continue to expand. New extremozymes from extremophiles are being identified and developed relatively slowly, since pharmaceutical industries are driven by economic gains from their innovations. However, increased investment for research into the characteristics of the various proteins and substances from extremophiles can help in the process of developing sustainable therapeutic solutions, as proposed in Fig. 4.1. Varying hypotheses regarding extremophiles' survival under harsh environmental conditions are still being explored. Therefore, there is a need for collaborative efforts to study extremophiles to resolve medical challenges in human society. Extremophiles have much untapped potential, and future research must investigate types of extremophiles that have not been previously studied for their possible uses in therapeutic and medical processes.

References

Abrevaya XC (2013) Features and applications of halophilic archaea. In: Singh OV (ed) Extremophiles—Sustainable resources and biotechnological implications. Wiley, Hoboken

Acharya S, Chaudhary A (2012) Bioprospecting thermophiles for cellulose production: a review. Braz J Microbiol 43:844–856

Achterman RR, White TC (2011) Dermatophyte Virulence factors: identifying and analyzing genes that may contribute to chronic or acute skin infections. Int J Microbiol 2012:358305

Adams MW, Kelly RM (1998) Finding and using hyperthermophilic enzymes. Trends Biotechnol 16:329–332

Adams MWW (1993) Enzymes and proteins from organisms that grow near and above 100 °C. Annu Rev Microbiol 47:58–627

Blumer-Schuette ES, Giannone RJ, Zurawski JV et al (2012) Caldicellulosiruptor core and pangenomes reveal determinants for noncellulosomal thermophilic deconstruction of plant biomass. J Bacteriol 194:4015–4028

Agrawala PK, Goel HC (2002) Protective effect of RH-3 with special reference to radiation induced micronuclei in mouse bone marrow. Indian J Exp Biol 40:525–530

Alexander B, Leach S, Ingledew WJ (1987) The relationship between chemiosmotic parameters and sensitivity to anions and organic acids in the acidophile *Thiobacillus ferrooxidans*. J Gen Microbiol 133:1171–1179

Almárcegui RJ, Navarro CA, Paradela A et al (2014) New copper resistance determinants in the extremophile *Acidithiobacillus ferrooxidans*: a quantitative proteomic analysis. J Proteome Res 13:946–960

Amaro AM, Chamorro D, Seeger M et al (1991) Effect of external pH perturbations on in vivo protein synthesis by the acidophilic bacterium *Thiobacillus ferrooxidans*. J Bacteriol 173:910–915

Arena A, Gugliandolo C, Stassi G et al (2009) An expolysaccharide produced by *Geobacillus thermodenitrificans* strain B3-72: antiviral activityon immunocompetent cells. Immunol Lett 123:132–137

Arora A, Ha C, Park CB (2004) Inhibition of insulin amyloid formation by small stress molecules. FEBS Lett 564:121–125

Asgarani E, Terato H, Asagoshi K et al (2000) Purification and characterization of a novel DNA repair enzyme from the extremely radioresistant bacterium *Rubrobacter radiotolerans*. J Radiat Res 41:19–34

Asker D, Awad TS, McLandsborough L et al (2011) *Deinococcus epolymerans* sp. nov., a gamma- and UV-radiation resistant bacterium, isolated from a radioactive site in Japan. Int J Syst Evol Microbiol 61:1448–14453

Baker-Austin C, Dopson M (2007) Life in acid: pH homeostasis in Acidophiles. Trends Microbiol 15:165–171

Barbara N, Giancula A, Annrita P (2013) Bacterial polymers produced by extremophiles; biosynthesis, characterization, and applications of expolysaccharides. In: Singh OV (ed) Extremophiles—Sustainable resources and biotechnological implications. Wiley, Hoboken

Bartlett DH, Chi E, Wright ME (1993) Sequence of the ompH gene from the deep-sea bacterium *Photobacterium* SS9. Gene 131:125–128

Batrakov SG, Pivovarova TA, Esipov SE et al (2002) Beta-D-glucopyranosyl caldarchaetidyl-glycerol is the main lipid of the acidophilic, mesophilic, ferrous iron-oxidising archaeon *Ferroplasma acidiphilum*. Biochim Biophys Acta 1581:29–35

Belfiore C, Ordoñez OF, Farías ME (2013) Proteomic approach of adaptive response to arsenic stress in *Exiguobacterium* sp. S17, an extremophile strain isolated from a high-altitude Andean Lake stromatolite. Extremophiles 17:421–431

Berger F, Morellet N, Menu F et al (1996) Cold shock and cold acclimation proteins in the psychrotrophic bacterium *Arthrobacter globiformis* SI55. J Bacteriol 178:2999–3007

Beyer N, Driller H, Bunger J (2000) Ectoin-a innovative, multi-functional active substance for the cosmetic industry. Seiden Ole Fette Wachse J 126:27–29

Booth IR (1985) Regulation of cytoplasmic pH in Bacteria. Microbiol Rev 49:359–378

Bordenstein S (2008) Microbial life in alkaline environments. http://www.serc.carleton.edu/ Accessed 20 Jul 2014

Bunger J, Driller H (2004) Ectoin: an effective natural substance to prevent UVA-induced premature photoaging. Skin Pharmacol Physiol 17:232–237

Buommino E, Schiraldi C, Baroni A et al (2005) Ectoine from halophilic microorganisms induces the expression of hsp70 and hsp70B' in human keratinocytes modulating the proinflammatory response. Cell Stress Chaperones 10:197–203

Cárdenas JP, Valdés J, Quatrini R et al (2010) Lessons from the genomes of extremely acidophilic bacteria and archaea with special emphasis on bioleaching microorganisms. Appl Microbiol Biotechnol 88:20–605

Castanie-Cornet MP, Penfound TA, Smith D et al (1999) Control of acid resistance in *Escherichia coli*. J Bacteriol 181:3525–3535

Cava F, Hidalgo A, Berenguer J (2009) Thermus thermophilus as biological model. Extremophiles 13:213–231

Cavicchioli R, Amils R, Wagner D et al (2011) Life and applications of extremophiles. Environ Microbiol 13:1903–1907

Chakravorty D, Patra S (2013) Attaining extremophiles and extremolytes: methodologies and limitations. In: Singh OV (ed) Extremophiles—Sustainable resources and biotechnological implications. Wiley, Hoboken, pp 29–74

Chi A, Valenzuela L, Beard S et al (2007) Periplasmic proteins of the extremophile *Acidithiobacillus ferrooxidans*: a high throughput proteomics analysis. Mol Cell Proteomics 6:2239–2251

Ciaramella M, Napoli A, Rossi M (2005) Another extreme genome: how to live at pH0. Trends Microbiol 13:49–51

Copeland E, Choy N, Gabani P et al (2013) Biosynthesis of extremolytes: radiation resistance and biotechnological implications. In: Singh OV (ed) Extremophiles—Sustainable resources and biotechnological implications. Wiley, Hoboken, pp 367–388

Corry B, Lee S, Ahern CA (2014) Pharmacological insights and quirks of bacterial sodium channels. Handb Exp Pharmacol 221:67–251

Craik CS, Page MJ, Madison EL (2011) Proteases as therapeutics. Biochem J 435:1–16

D'Amico S, Collins T, Marx JC et al (2006) Psychrophilic microorganisms: challenges for life. EMBO Rep 7:385–389

Dahl MV (1987) Immunological resistance to dermatophyte infections. Adv Dermatol 2:305–320

Danson MJ, Hough DW (1998) Structure, function and stability of enzymes from the Archaea. Trends Microbiol 6:307–314 de la Coba F, Aguilera J, de Galvez MV et al (2009) Prevention of the ultraviolet effects on clinical and histopathological changes, as well as the heat shock protein-70 expression in mouse skin by topical applications of algal UV-absorbing compounds. J Dermatol Sci 55:161–169

De Maayer P, Anderson D, Cary C et al (2014) Some like it cold: understanding the survival strategies of psychrophiles. EMBO Rep 15:508–517

de Vrese M, Kristen H, Rautenberg P et al (2011) Probiotic lactobacilli and bifidobacteria in a fermented milk product with added fruit preparation reduce antibiotic associated diarrhea and *Helicobacter pylori* activity. J Dairy Res 78:396–403

Demirjian DC, Moris-Varas F, Cassidy CS (2001) Enzymes from extremophiles. Curr Opin Chem Biol 5:144–151

Duan Z, Ji D, Weinstein EJ et al (2010) Lentiviral shRNA screen of human kinases identifies PLK1 as a potential therapeutic target for osteosarcoma. Cancer Lett 293:220–229

Duplantis BN, Osusky M, Schmerk CL et al (2010) Essential genes from Arctic bacteria used to construct stable, temperature-sensitive bacterial vaccines. Proc Natl Acad Sci USA 2107:13456–13460

Edwards KJ, Bond PL, Gihring TM et al (2000) An archaeal Iron-oxidizing extreme acidophile important in mine drainage. Science 287:1796–1799

Egorova K, Antranikian G (2005) Industrial relevance of thermophilic Archaea. Curr Opin Microbiol 8:649–655

Eichler J (2001) Biotechnological uses of archaeal extremozymes. Biotechnol Adv 19:261–278

Elleuche S, Piascheck H, Antranikian G (2011) Fusion of the OsmC domain from esterase EstO confers thermolability to the cold-active xylanase Xyn8 from *Pseudoalteromonas arctica*. Extremophiles 15:311–317

Elleuche S, Schröder C, Sahm K et al (2014) Extremozymes-biocatalysts with unique properties from extremophilic microorganisms. Curr Opin Biotechnol 29C:116–123

Elmer GW (2001) Probiotics: "living drugs". Am J Health Syst Pharm 58:1101–1109

Emiley A, Malfatti EF, Gutierrez J et al (2011) Isolation and characterization of a psychropiezophilic alphaproteobacterium. Appl Environ Microbiol 77:8145–8153

Espina G, Eley K, Pompidor G et al (2014) A novel β-xylosidase structure from *Geobacillus thermoglucosidasius*: the first crystal structure of a glycoside hydrolase family GH52 enzyme reveals unpredicted similarity to other glycoside hydrolase folds. Acta Crystallogr D Biol Crystallogr 70:1366–1374

Feller G (2003) Molecular adaptations to cold in psychrophilic enzymes. Cellular Mol Life Sci 60:648–662

Feller G, Gerdey C (2003) Psychrophilic enzymes: hot topics in cold adaptation. Nature 1:200–208

Ferroni L, Klisch M, Pancaldi S et al (2010) Complementary UV-absorption of mycosporine-like amino acids and scytonemin is responsible for the UV-insensitivity of photosynthesis in *Nostoc flagelliforme*. Mar Drugs 8:106–121

Fornbacke M, Clarsund M (2013) Cold-adapted proteases as an emerging class of therapeutics. Infect Dis Ther 2:15–26

Foster JW (2004) *Echerichia coli* acid resistance: tales of an amateur acidophile. Nat Rev 2:898–907

Furusho K, Yoshizawa T, Shoji S (2005) Ectoine alters subcellular localization of inclusions and reduces apoptotic cell death induced by the truncated Machado-Joseph disease gene product with an expanded polyglutamine stretch. Neurobiol Dis 20:170–178

Gabani P, Copeland E, Chandel AK et al (2012) Ultraviolet-radiation-resistant isolates revealed cellulose-degrading species of *Cellulosimicrobium cellulans* (UVP1) and *Bacillus pumilus* (UVP4). Biotechnol Appl Biochem 59:395–404

Gabani P, Prakash D, Singh OV (2014) Bio-signature of Ultraviolet-radiation-resistant extremophiles from elevated land. Am J Microbiol Res 2:94–104

Gabani P, Singh OV (2013) Radiation-resistant extremophiles and their potential in biotechnology and therapeutics. Appl Microbiol Biotechnol 97:993–1004

Garcia-Descalzo L, Alcazar A, Baquero F et al (2013) Biotechnological applications of cold-adapted bacteria. In: Singh OV (ed) Extremophiles—Sustainable resources and biotechnological implications. Wiley, Hoboken, pp 159–174

Georlette D, Damien B, Blaise V et al (2003) Structural and functional adaptations to extreme temperatures in psychrophilic, mesophilic and thermophilic DNA ligases. J Biol Chem 278:37015–37023

Golyshina OV, Pivovarova TA, Karavaiko GI et al (2000) *Ferroplasma acidiphilum* gen. nov., sp. nov., an acidophilic, autotrophic, ferrous-iron-oxidizing, cell-wall lacking, mesophilic member of the erroplasmaceae fam. nov., comprising a distinct lineage of the Archaea. Int J Syst Evol Microbiol 50:997–1006

Gotteland M, Andrews M, Toledo M et al (2008) Modulation of Helicobacter pylori colonization with cranberry juice and *Lactobacillus johnsonii* La1 in children. Nutrition 24:421–426

Grandy G, Medina M, Soria R et al (2010) Probiotics in the treatment of acute rotavirus diarrhoea. A randomized, double-blind, controlled trial using two different probiotic preparations in Bolivian children. BMC Infect Dis 10:253

Gudmundsdottir A, Palsdottir HM (2005) Atlantic cod trypsins: from basic research to practical applications. Mar Biotechnol 7:77–88

Guiliani N, Jerez CA (2000) Molecular cloning, sequencing, and expression of omp-40, the gene coding for the major outer membrane protein from the acidophilic bacterium *Thiobacillus ferrooxidans*. Appl Environ Microbiol 66:2318–2324

Guo X, Yin H, Liang Y et al (2014) Comparative genome analysis reveals metabolic versatility and environmental adaptations of *Sulfobacillus thermosulfidooxidans* strain ST. PLoS One 9: e99417

Hendry P (2006) Extremophiles: there's more to life. Environ Chem 3:75–76

Hieter P, Boguski M (1997) Functional genomics: it's all how you read it. Science 278:601–602

Horikoshi K (1999) Alkaliphiles: some applications of their products for biotechnology. Microbiol Mol Biol Rev 63:735–750

Horikoshi K, Grant WD (1998) Extremophiles microbial life in extreme environments. Wiley, New York

Hough DW, Danson MJ (1999) Extremozymes. Curr Opin Chem Biol 3:39–46

Insel G, Celikyilmaz G, Ucisik-Akkaya E et al (2006) Respirometric evaluation and modeling of glucose utilization by *Escherichia coli* under aerobic and mesophilic cultivation conditions. Biotechnol Bioeng 96:94–105

Irgens RL, Gosink JJ, Staley JT (1996) *Polaromonas vacuolata* gen. nov., sp. nov., a psychrophilic, marine, gas vacuolate bacterium from Antarctica. Int J Syst Bacteriol 46:822–826

Irwin JA, Baird AW (2004) Extremophiles and their application to veterinary medicine. Ir Vet J 57:348–354

Ishii N, Nakahigashi K, Baba T et al (2007) Multiple high-throughput analyses monitor the response of *E. coli* to perturbations. Science 316:593–597

Jamil F, Teh AH, Schadich E et al (2014) Crystal structure of truncated haemoglobin from an extremely thermophilic and acidophilic bacterium. J Biochem 156:97–106

Jaubert C, Danioux C, Oberto J et al (2013) Genomics and genetics of *Sulfolobus islandicus* LAL14/1, a model hyperthermophilic archaeon. Open Biol 3:130010

Jiang CY, Liu LJ, Guo X et al (2014) Resolution of carbon metabolism and sulfur-oxidation pathways of *Metallosphaera cuprina* Ar-4 via comparative proteomics. J Proteomics. doi:10.1016/j.jprot.2014.07.004

Jorge CD, Ventura R, Maycock C et al (2011) Assessment of the Efficacy of Solutes from extremophiles on Protein aggregation in cell models of Huntington's and Parkinson's diseases. Neurochem Res 36:1005–1011

Kambourova M, Mandeva R, Dimova D et al (2009) Production and characterization of a microbial glucan, synthesized by *Geobacillus tepidamans* V264 isolated from Bulgarian hot spring. Carbohydr Poly 77:338–343

Kanapathipillai M, Lentzen G, Sierks M et al (2005) Ecotine and hydroxyecotine inhibit aggregation and neurotoxicity of Alzheimer's β-amyloid. FEBS Lett 579:4775–4780

Karan R, Capes MD, Dassarma S (2012) Function and biotechnology of extremophilic enzymes in low water activity. Aquat Biosyst 8:4

Karsten U, Escoubeyrou K, Charles F (2009) The effect of redissolution solvents and HPLC columns on the analysis of mycosporine-like amino acids in the eulittoral macroalgae Prasiolacrispa and Porphyraumbilicalis. Hegol Mar Res 63:231–238

Kato C, Bartlett DH (1997) The molecular biology of barophilic bacteria. Extremophiles. 1:111–116

Kato C, Inoue A, Horikoshi K (1996) Isolating and characterizing deep-sea marine microorganisms. Trends Biotechnol 14:6–12

Kato C, Masui N, Horikoshi K (1996) Properties of obligately barophilic bacteria isolated from a sample of deep-sea sediment from the Izu-Bonin trench. J Mar Biotechnol 4:96–99

Kato C, Sato T, Horikoshi K (1995) Isolation and properties of barophilic and barotolerant bacteria from deep-sea mud samples. Biodivers Conserv 4:1–9

Kishimoto N, Inagaki K, Sugio T et al (1990) Growth-inhibition of Acidiphilium species by organic-acids contained in yeast extract. J Ferment Bioeng 70:7–10

Kobayashi T, Lu J, Li Z et al (2007) Extremely high alkaline protease from a deep-subsurface bacterium, *Alkaliphilus transvaalensis*. Appl Microbiol Biotechnol 75:71–80

Kohama Y, Tanaka K, Takae S et al (1994) Studies on Thermophile products. VII Effect of 1,3-Di-14-methylpentadecanoyl Glycerol and its related Isofatty acids on T cell proliferation in vitro. Biol Pharm Bulletin 17:850–852

Konings WN, Albers SV, Koning S et al (2002) The cell membrane plays a crucial role in survival of bacteria and archaea in extreme environments. Antonie van Leeuwenhoek 81:61–72

Kort JC, Esser D, Pham TK et al (2013) A cool tool for hot and sour Archaea: proteomics of *Sulfolobus solfataricus*. Proteomics 13:2831–2850

Ksouri R, Ksouri WM, Jallali I et al (2012) Medicinal halophytes: potent source of health promoting biomolecules with medical, nutraceutical and food applications. Crit Rev Biotechnol 32:289–326

Kufner K, Lipps G (2013) Construction of a chimeric thermoacidophilic beta-endoglucanase. BMC Biochem 14:11

Kumar R, Patel DD, Bansal DD et al (2010) Extremophiles: sustainable resource of natural compound-Extremolytes. In: Singh OV, Harvey SP (eds) Sustainable biotechnology: sources of renewable energy. Springer Press, UK, pp 279–294

Kumar R, Singh A (2013) Smart therapeutics from extremophiles: unexplored applications and technological challenges. In: Singh OV (ed) Extremophiles—Sustainable resources and biotechnological implications. Wiley, Hoboken, pp 389–401

Lakshmipathy D, Kannabiran K (2010) Review on dermatomycosis: pathogenesis and treatment. J Nat Sci 2:726–731

Lauro FM, Bartlett DH (2007) Prokaryotic lifestyles in deep sea habitats. Extremophiles 12:15–25

Le Bail JC, Varnat F, Nicolas JC et al (1998) Estrogenic and antiproliferative activities on MCF-7 human breast cancer cells by flavonoids. Cancer Lett 130:16–209

Lee AM, Sevinsky JR, Bundy JL et al (2009) Proteomics of *Pyrococcus furiosus*, a hyperthermophilic archaeon refractory to traditional methods. J Proteome Res 8:3844–3851

Lee YP, Giorgi FM, Lohse et al (2013) Transcriptome sequencing and microarray design for functional genomics in the extremophile Arabidopsis relative Thellungiella salsuginea (Eutrema salsugineum). BMC Genomics 14:793

Lentzen G, Schwarz T (2006) Extremolytes: natural compounds from extremophiles for versatile applications. Appl Microbiol Biotechnol 72:623–634

Liévin-Le MV, Sarrazin-Davila LE, Servin AL (2007) An experimental study and a randomized, double-blind, placebo-controlled clinical trial to evaluate the antisecretory activity of *Lactobacillus acidophilus* strain LB against nonrotavirus diarrhea. Pediatrics 120:e795–e803

Liljeqvist M, Rzhepishevska OI, Dopson M (2013) Gene identification and substrate regulation provide insights into sulfur accumulation during bioleaching with the psychrotolerant acidophile *Acidithiobacillus ferrivorans*. Appl Environ Microbiol 79:951–957

Lin L, Xu J (2013) Dissecting and engineering metabolic and regulatory networks of thermophilic bacteria for biofuel production. Biotechnol Adv 31:37–827

Lin PP, Rabe KS, Takasumi JL et al (2014) Isobutanol production at elevated temperatures in thermophilic *Geobacillus thermoglucosidasius*. Metab Eng 24:1–8

Liu S, Zhang P, Cong B et al (2010) Molecular cloning and expression analysis of a cytosolic Hsp70 gene from Antarctic ice algae *Chlamydomonas* sp. ICE-L Extremophiles 14:329–337

Liu X, Huang Z, Zhang X et al (2014) Clonig, expression, and characterization of a novel cold-active and halophilic xylanase from *Zunongwangia profunda*. Extremophiles 18:441–450

Liu Z, Zhao X, Bai F (2013) Production of xylanase by an alkaline-tolerant marine-derived Streptomyces viridochromogenes strain and improvement by ribosome engineering. Appl Microbiol Biotechnol 97:4361–4368

Llamas I, Béjar V, Martínez-Checa F et al (2011) Halomonas stenophila sp. nov., a halophilic bacterium that produces sulphate exopolysaccharides with biological activity. Int J Syst Evol Microbiol 61:2508–2514

Llewellyn CA, Airs RL (2010) Distribution and abundance of MAAs in 33 species of microalgae across 13 classes. Mar Drugs 8:1273–1291

Lopez FJ, Gomez MJ, Gonzalez E et al (2013) Dyn Genomes Acidophiles Polyextremophiles 27:81–97

Lu S, Gischkat S, Reiche M et al (2010) Ecophysiology of Fe-Cycling bacteria in acidic sediments. Appl Environ Microbiol 76:8174–8183

Luo J, Emanuele MJ, Li D et al (2009) A genome-wide RNAi screen identifies multiple synthetic lethal interactions with the Ras oncogene. Cell 137:825–848

Ma Y, Galinski E, Grant WD et al (2010) Halophiles 2010: Life in saline environments. Appl Environ Microbiol 76:6971–6981

Macalady J, Banfield JF (2003) Molecular geomicrobiology: genes and geochemical cycling. Earth Planet Sci Lett 209:1–17

MacElroy M (1974) Some comments on the evolution of extremophiles. Biosystems 6:74–75

Majhi MC, Behera AK, Kulshreshtha NM et al (2013) ExtremeDB: a unified web repository of extremophilic archaea and bacteria. PLoS One 8:e63083

Mallik S, Kundu S (2014) Molecular interactions within the halophilic, thermophilic, and mesophilic prokaryotic ribosomal complexes: clues to environmental adaptation. J Biomol Struct Dyn. doi:10.1080/07391102.2014.900457

Mangold S, Valdés J, Holmes DS, Dopson M (2011) Sulfur metabolism in the extreme acidophile Acidithiobacillus caldus. Front Microbiol 2:17. doi:10.3389/fmicb.2011.00017. eCollection 2011

Mantelli F, Scala C, Ronchi A et al (2003) Macrocostituenti de elementi in traccia nelle acque dei laghi saline delle Andi de Catamarca e La Rioja (Argentina). Boll Chim Igien 54:37–44. (: Cross reference)

Margesin R, Schinner F (1994) Properties of cold-adapted microorganisms and their potential role in biotechnology. J Biotechnol 33:1–14

Marquis RE, Keller DM (1975) Enzymatic adaptation by bacteria under pressure. J Bacteriol 122:575–584

Marteinsson VT, Birrien JL, Reysenbach AL et al (1999) *Thermococcus barophilus* sp. nov., a new barophilic and hyperthermophilic archaeon isolated under high hydrostatic pressure from a deep-sea hydrothermal vent. Int J Sys Bacteriol 49:351–359

Marx JC, Collins T, D'Amico S et al (2007) Cold-adapted enzymes from marine Antarctic microorganisms. Mar Biotechnol 9:293–304

Matin A (1990) Keeping a neutral cytoplasm; the bioenergetics of obligate acidophiles. FEMS Microbiol Rev 75:307–318

Merkley ED, Daggett V, Parson WW (2011) A temperature-dependent conformational change of NADH oxidase from *Thermus thermophilus* HB8. Proteins. doi:10.1002/prot.23219

Metpally R, Reddy B (2009) Comparative proteome analysis of psychrophilic versus mesophilic bacterial species: insights into the molecular basis of cold adaptation of proteins. BMC Genomics 10:1–10

Michels M, Bakker EP (1985) Generation of a large, protonophore-sensitive proton motive force and pH difference in the acidophilic bacteria Thermoplasma acidophilum and *Bacillus acidocaldarius*. J Bacteriol 161:231–237

Michetti P, Dorta G, Wiesel PH et al (1999) Effect of whey-based culture supernatant of *Lactobacillus acidophilus* (johnsonii) La1 on *Helicobacter pylori* infection in humans. Digestion 60:203–209

Molina IJ, Ruiz-Ruiz C (2013) Biomedical applications of exopolysaccharides produce by microorganisms isolated from extreme environments. In: Singh OV (ed) Extremophiles— Sustainable resources and biotechnological implications. Wiley, Hoboken, pp 357–366

Morimoto RI, Santoro MG (1998) Stress-inducible responses and heat shock proteins: new pharmacologic targets for cytoprotection. Nat Biotechnol 16:8–833

Mukherjee P, Fulton DB, Halder M et al (2006) Maristentorin, a novel pigment from the positively phototactic marine ciliate *Maristentor dinoferus*, is structurally related to hypericin and stentorian. J Phys Chem B 110:6359–6364

Muller K (2001) Pharmaceutically relevant metabolites from lichens. Appl Microbiol Biotechnol 56:9–16

Nakasone K, Kato C, Horikoshi K (1996) Molecular Cloning of the gene encoding RNA polymerase α subunit from deep-sea barophilic bacterium. Biochim Biophys Acta 1308:107–110

Nakayama H, Yoshida K, Ono H et al (2000) Ectoine, the compatible solute of *Halomonas elongate*, confers hyperosmotic tolerance in cultured tobacco cells. Plant Physiol 122:1239–1247

Niehaus F, Bertoldo C, Kahler M et al (1999) Extremophiles as a source of novel enzymes for industrial application. Appl Microbiol Biotechnol 51:711–729

Oost J, Antranikian G (1996) Extremophiles. Tibtech. Forum 14:415–417

Oren A (2002) Diversity of halophilic micororganisms: environment, phylogeny, physiology, and applications. J Ind Microbiol Biotechnol 28:56–63

Ortenberg R, Rozenblatt-Rosen O, Mevarech M (2000) The extremely halophilic archaeon *Haloferax volcanii* has two very different dihydrofolate reductases. Mol Microbiol 35:1493–1505

Osorio H, Mangold S, Denis Y et al (2013) Anaerobic sulfur metabolism coupled to dissimilatory iron reduction in the extremophile *Acidithiobacillus ferrooxidans*. Appl Environ Microbiol 79:2172–2181

Osorio H, Martínez V, Nieto PA et al (2008) Microbial iron management mechanisms in extremely acidic environments: comparative genomics evidence for diversity and versatility. BMC Microbiol 8:203

Ouwehand AC, ten Bruggencate SJ, Schonewille AJ et al (2014) *Lactobacillus acidophilus* supplementation in human subjects and their resistance to enterotoxigenic *Escherichia coli* infection. Br J Nutr 111:465–473

Oyamada C, Kaneniwa M, Ebitani K et al (2008) Mycosporine-like amino acid extracted from scallop (Patinopectenyessoensis) ovaries, UV protection and growth stimulation activities on human cells. Mar Biotechnol 10:141–150

Pantoflickova D, Corthesy-Theulaz I, Dorta G et al (2003) Favourable effect of regular intake of fermented milk containing *Lactobacillus johnsonii* on *Helicobacter pylori* associated gastritis. Aliment Pharmacol Ther 18:805–814

Pennacchia C, Breeuwer P, Meyer R (2014) Development of a Multiplex-PCR assay for the rapid identification of *Geobacillus stearothermophilus* and *Anoxybacillus flavithermus*. Food Microbiol 43:41–49

Pereira SL, Reeve JN (1998) Histones and nucleosomes in archaea and eukarya: A comparative analysis. Extremophiles 2:141–148

Piette F, Leprince P, Feller G (2012) Is there a cold shock response in the Antarctic psychrophile *Pseudoalteromonas haloplanktis*? Extremophiles 16:681–683

Pivovarova TA, Kondrat'eva TF, Batrakov SG et al (2002) Phenotypic features of *Ferroplasma acidiphilum* strains Y-T and Y-2. Mikrobiologiia 71:809–818

Qin Y, Huang Z, Liu Z (2014) A novel cold-active and salt-tolerant α-amylase from marine bacterium *Zunongwangia profunda*: molecular cloning, heterologous expression and biochemical characterization. Extremophiles 18:271–281

Rastogi RP, Richa SRP, Singh SP et al (2010) Photoprotective compounds from marine organisms. J Ind Microbiol Biotechnol 37:537–558

Rothschild LJ (2007) Definig the envelope for the search for life in the universe. In: Pudritz RE (ed) Planetarysystemsand the origin of life. Cambridge University Press, NY

Rothschild LJ, Mancinelli RL (2001) Life in extreme environments. Nature 409:1092–1101

Ruiz-Ruiz C, Srivastava GK, Carranza D et al (2011) An exopolysaccharide produced by the novel halophilic bacterium *Halomonas stenophila* strain B100 selectively induces apoptosis in human T leukaemia cells. Appl Microbiol Biotechnol 89:345–355

Russo A, Piovano M, Lombardo L et al (2008) Lichen metabolites prevent UV light and nitric oxide-mediated plasmid DNA damage and induce apoptosis in human melanoma cells. Life Sci 83:468–474

Saaranen MJ, Ruddock LW (2013) Disulfide bond formation in the cytoplasm. Antioxid Redox Signal 19:46–53

Sachs G, Scott DR, Wen Y (2011) Gastric Infection by Helicobacter pylori. Curr Gastroenterol Rep 11:455–461

Saelensminde G, Halskau O (2008) Amino acid contacts in proteins adapted to different temperatures: hydrophobic interactions and surface charges play a key role. Extremophiles. 13:11–20

Sellek GA, Chaudhuri JB (1999) Biocatalysis in organic media using enzymes from extremophiles. Enzy Microbial Technol 25:471–482

Shahmohammadi HR, Asgarani E, Terato H et al (1998) Protective roles of bacteriotuberin and intracellular KCl in the resistance of *Halobacterium salinarium* against DNA damaging agents. J Radiat Res 39:251–262

Shanmugam MM, Parasuraman S (2012) Evolutionary conserved essential genes from arctic bacteria: a tool for vaccination. J Young Pharmacists 4:55–57

Sharma A, Kawarabayasi Y, Satyanarayana T (2012) Acidophilic bacteria and archaea: acid stable biocatalysts and their potential applications. Extremophiles 16:1–19

Shimada H, Nemoto N, Shida Y et al (2002) Complete polar lipid composition of *Thermoplasma acidophilum* HO-62 determined by high-performance liquid chromatography with evaporative light-scattering detection. J Bacteriol 184:556–563

Shin DS, Pratt AJ, Tainer JA (2014) Archaeal genome guardians give insights into eukaryotic DNA replication and damage response proteins. Archaea 2014:206735

Siddiqui MA, Rashid N, Ayyampalayam S et al (2014) Draft genome sequence of *Geobacillus thermopakistaniensis* strain MAS1. Genome Announc 2.e00559–e00614

Simon MM, Reikerstorfer A, Schwarz A et al (1995) Heat shock protein 70 overexpression affects the response to ultraviolet light in murine fibroblasts. Evidence for increased cell viability and suppression of cytokine release. J Clin Invest 95:926–933

Singh AK, Pindi PK, Dube S et al (2009) Importance of trmE for Growth of the psychrophile *Pseudomonas syringae* at low temperatures. Appl Environ Microbiol 75:4419–4426

Singh OV (2006) Proteomics and metabolomics, the molecular make-up of toxic aromatic pollutant bioremediation. Proteomics 6:5481–5492

Singh OV (2013) (ed) Extremophiles—Sustainable resources and biotechnological implications. Wiley, Hoboken

Singh OV, Gabani P (2011) Extremophiles: radiation resistance microbial reserves and therapeutic implications. J Appl Microbiol 110:851–861

Soule T, Palmer K, Gao Q et al (2009) A comparative genomics approach to understanding the biosynthesis of sunscreen scytonemin in cyanobacteria. BMC Genomics 10:336–346

Stevenson CS, Capper EA, Roshak AK et al (2002) The identification and characterization of the marine natural product scytonemin as a novel antiproliferative pharmacophore. J Pharmacol Exp Ther 303:858–866

Stevenson CS, Capper EA, Roshak AK et al (2002) Scytonemin, a marine natural product inhibitor of kinases key in hyperproliferative inflammatory diseases. Inflamm Res 51:112–114

Sun J, Shen P, Chao H et al (2009) Isolation and identification of a new radiation-resistant bacterium *Deinococcus guangrensis* sp. nov. and analysis of its radioresistant character. Wei Sheng Wu Xue Bao 49:918–924

Suzuki H, Yoshida K, Ohshima T (2013) Polysaccharide-degrading thermophiles generated by heterologous gene expression in *Geobaccillus kaustophilus* HTA426. Appl Environ Microbiol 79:5151–5158

Taguchi S, Ozaki A, Momose H (1998) Engineering of a cold-adapted protease by sequential random mutagenesis and a screening system. Appl Environ Microbiol 64:492–495

Takai K, Moser DP, Onstott TC et al (2001) *Alkaliphilus transvaalensis* gen. nov., sp. nov., an extremely alkaliphilic bacterium isolated from a deep South African gold mine. Int J Syst Evol Microbiol 51:1245–1256

Tan JCH, Kalapesi FB, Coroneo MT (2006) Mechanosensitivity and the eye: cells coping with the pressure. J Opthomol 90:383–388

Thangakani AM, Kumar S, Velmurugan D et al (2012) How do thermophilic proteins resist aggregation? Proteins 80:1003–1015

Thanos D, Maniatis T (1995) Identification of the rel family members required for virus induction of the human beta interferon gene. Mol Cell Biol 15:152–164

Tomlinson GA, Strohm MP, Hochstein LI (1978) The metabolism of carbohydrates by extremely halophilic bacteria: the identification of lactobionic acid as a product of lactose metabolism by *Halobacterium saccharovorum*. Can J Microbiol 24:898–903

Torres A, Enk CD, Hochberg M et al (2006) Porphyra-334, a potential natural source for UVA protective sunscreens. Photochem Photobiol Sci 5:432–435

Tyson GW, Chapman J, Hugenholtz P et al (2004) Community structure and metabolism through reconstruction of microbial genomes from the environment. Nature 428:37–43

van de Vossenberg JL, Driessen AJ, Konings WN (1998) The essence of being extremophilic: the role of the unique archaeal membrane lipids. Extremophiles 2:163–170

van de Vossenberg JL, Driessen AJ, Zilling et al (1998b) Bio-energetics and cytoplasmic membrane stability of the extremely acidophilic, thermophilic archaeon, Picrophilus oshimae. Extremophiles 2:67–74

van Wolferen M, Ajon M, Driessen AJ et al (2013) How hyperthermophiles adapt to change their lives: DNA exchange in extreme conditions. Extremophiles 17:545–563

Wang W, Mao J, Zhang Z et al (2009) *Deinococcus wulumuqiensis* sp. nov., and *Deinococcus xibeiensis* sp. nov., isolated from radiation-polluted soil. Int J Syst Evol Microbiol 60:2006–2010

Weitzman I, Summerbell RC (1995) The Dermatophytes. Clin Microbiol Rev 8:240–259

Welch WJ (1993) Heat shock proteins functioning as molecular chaperones: their roles in normal and stressed cells. Philos Trans R Soc Lond B Biol Sci 339:327–333

Wemheuer B, Taube R, Akyol P et al (2013) Microbial diversity and biochemical potential encoded by thermal spring metagenomes derived from the Kamchatka Peninsula. Archaea. 2013:136714

Widdel F, Bak F (1992) Gram-negative mesophilic sulfate-reducing bacteria. In: Balows A, Trüper HG, Dworkin M, Harder W, Schleifer KH (eds) The Prokaryotes, 2nd edn. Springer, New York, pp 3352–3378

Woappi Y, Gabani P, Singh A et al (2014) Antibiotrophs: The complexity of antibiotic-subsisting and antibiotic-resistant microorganisms. Crit Rev Microbiol. doi:10.3109/1040841X.2013.875982

Yamauchi S, Ueda Y, Matsumoto M et al (2012) Distinct features of protein folding by the GroEL system from a psychrophilic bacterium, *Colwellia psychrerythraea* 34H. Extremophiles. 16:871–882

Yano Y, Nakayama A, Ishihara K et al (1998) Adaptive changes in membrane lipids of barophilic bacteria in response to changes in growth pressure. Appl Environ Microbiol 64:479–485

Yao N, Ren Y, Wang W (2013) Genome Sequence of a thermophilic *Bacillus, Geobacillus thermodenitrificans* DSM465. Genome Announc (1pii:e01046-13)

Yoo CG, Lee S, Lee CT et al (2000) Anti-inflammatory effect of heat shock protein induction is related to stabilization of IκBα through preventing IκB kinase activation in respiratory epithelial cells. J Immunol 164:5416–5423

Yuan M, Zhang W, Dai S et al (2009) *Deinococcus gobiensis* sp. nov., an extremely radiation-resistant bacterium. Int J Syst Evol Microbiol 59:1513–1517

Yuan YV, Westcott ND, Kitts DD et al (2009) Mycosporine-like amino acid composition of the edible red alga, *Palmaria palmate* (dulse) harvested from the west and east costs of Grand Manan Island, New Brunswick. Food Chem 112:321–328

Yun NR, Lee YN (2009) Iso-superoxide dismutase in *Deinococcus grandis*, a UV resistant bacterium. J Microbiol 47:172–177

Zecchinon L, Claverie P, Collins T et al (2001) Did psychrophilic enzymes really win the challenge? Extremophiles 5:313–321

Zhang L, Li L, Wu Q (2007) Protective effects of mycosporine-like amino acids of *Synechocystis* sp. PCC 6803 and their partial characterization. J Photochem Photobiol B 86:240–245

Zheng H, Wu H (2010) Gene-centric association analysis for the correlation between the guanine-cytosine content levels and temperature range conditions of prokaryotic species. BMC Bioinformatics 14(11 Suppl 11):S7

Zhou X, Tron VA, Li G et al (1998) Heat shock transcription factor-1 regulates heat shock protein-72 expression in human keratinocytes exposed to ultraviolet B light. J Investigative Dermatology 111:194–198

Zügel U, Kaufmann SHE (1999) Role of heat shock proteins in protection from and pathogenesis of infectious diseases. Clin Microbiol Rev 12:19–39

Zychlinsky E, Matin A (1983) Cytoplasmic pH homeostasis in an acidophilic bacterium, *Thiobacillus acidophilus*. J Bacteriol 156:1352–1355